HOME ROBOTICS

First published in Great Britain in 2018
by Aurum Press an imprint of
The Quarto Group
The Old Brewery
6 Blundell Street London
N7 9BH

www.QuartoKnows.com

A catalogue record for this book is available from
the British Library.

Edited by Philip de Ste. Croix
Photography by Daniel Knox
Designed and produced by Sue Pressley and
    Paul Turner, Stonecastle Graphics Ltd

The author would like to thank Michael Berry
and Martin Faulkner for proofreading, Jutta Knox
for her patience during all the late nights that
he spent working in the garage, and his nieces
Emily and Isabel Phillips for all their great
robot ideas.

ISBN 978-1-78131-700-6

1 3 5 7 9 10 8 6 4 2

2018 2020 2022 2021 2019

Printed in China

HOME
ROBOTICS

MAKER-INSPIRED PROJECTS FOR
BUILDING YOUR OWN ROBOTS

DANIEL KNOX

A

Everyone can enjoy building robots – it's easy and great fun! Start with our **BASIC BOTS** that use simple components, progress to the more advanced **SIMPLE ROBOTS** controlled by the BBC micro:bit and with practice, you'll soon be building the great **SMART MAKES** with confidence.

CONTENTS

INTRODUCTION	6
TOOLS GUIDE	10
COMPONENTS GUIDE	12
BASIC TECHNIQUES	15

01

BASIC BOTS

PAGE 16

BRISTLE BOT — 18
A battery-powered vibrating motor transforms an everyday nail brush into a crazy machine that will skitter energetically around on your table top.

SQUIBBLE BOT — 24
By attaching a couple of marker pens to an upturned drinks cup, we can make a bot that produces mind-blowing scribbles of its own accord.

BUTTERFLY BOT — 30
Building on the techniques we learned in Bristle Bot, two toothbrush heads are turned into a space-age butterfly that skips along on its own.

ROBO ROACH — 36
Robo Roach is equipped with insect-like feeler sensors that alert it when it encounters an obstacle and cause it to steer away from the obstruction.

SPIRO BOT — 48
This robot brings the Spirograph toy into the 21st century as we build a machine that creates psychedelic patterns of amazing variety.

SIMPLE ROBOTS

PAGE 56

AVATAR 58
This project introduces you to computer programming using the JavaScript Blocks Editor to control the display on a BBC micro:bit

SCUTTLE BOT 66
Scuttle Bot transforms a metal tin into a machine that moves around under the control of a program on your mobile phone.

CATAPULT BOT 78
The Catapult Bot uses an infra-red motion sensor to activate the catapult's firing mechanism when movement is detected.

GARDEN GUARDIAN 88
Robotics comes to your aid with this brilliant little machine that alerts your phone when a pot plant needs to be watered.

WALKING ROBOT 98
This robot tackles a considerable technical challenge – how to walk on two legs – and pulls it off with an ingenious solution.

SMART MAKES

PAGE 110

ROBO WARRIOR 112
Fighting robots don't have to be massively complicated to be effective in the combat zone – this lightweight warrior packs a mighty punch.

CNC WRITER 126
This project introduces readers to the world of computer numerical control as we build a cool machine that can write messages on a pad.

MARS ROVER 142
NASA sent one to explore the surface of Mars, and here we learn how to construct an intelligent all-terrain rover vehicle along the same lines.

RESOURCES 158

INDEX 159

INTRODUCTION

Building robots can at first seem like an impossible challenge, but really the only skill that you must have is a passion for making things – everything else is just a matter of learning and practice.

In this book I will show you how to build a number of different robots; some can walk, others can draw and a few are inspired by machines that help humans every day. As you build these simple machines, you will learn to put together electronic circuits, write pieces of software and use various tools to assemble the bots. Who knows, once you have got a taste for it, you may develop into one of our star robotics engineers?

01 BASIC BOTS

Every roboticist must start somewhere and there's no better place than using what's readily available around the home. This first chapter features a series of simple mechanical robots that will hone your basic skills. These projects include Bristle Bot, Squibble Bot and Butterfly Bot – a trio of vibration robots that can skitter about. Robo Roach uses a readily available chassis that can be adapted to create an autonomous touch-sensitive robot that finds its own way around, while Spiro Bot can create amazing drawings on its own.

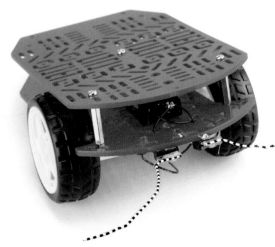

02 SIMPLE ROBOTS

Many robots use small electronic devices to imbue them with intelligence and precision control. Programming can at first seem a little scary, but in reality it's just thinking about the world in an ordered, systematic way. To help you take your first steps into the world of software, I will teach you a graphical programming language that is simple to learn. Every piece of software also needs something to run on and so I have selected the BBC micro:bit as our embedded platform of choice. It's extremely simple to upload new programs to and has a lot of cool sensors built in. With these simple programs and the micro:bit, we will build a variety of intelligent machines including Walking Robot and Scuttle Bot that can sense and interact with their environment.

03 SMART MAKES

Once you have mastered the earlier projects, you will probably want to take on something a bit more complicated. The three advanced robots in this book have been designed to give you the flexibility to expand and modify them as you see fit. They do require a few more tools and parts than the other robots, but don't worry – there is a simple step-by-step guide to help you build them.

The first of them is a Robo Warrior designed for combat. Fighting with small robotic machines is a popular competitive sport that is enjoyed around the world. There are a few important rules to follow, but don't worry – I will teach you all you need to know and show you how to build a simple machine that can be easily adapted and customized. If you have access to a laser cutter, I have designed a template to help you get started with your combat bot. If you don't have access to one, don't worry – stiff cardboard works fine too. I have also created a file complete with a step-by-step guide to help you configure your transmitter if you want to change the way the robot is steered. All the files can be found and downloaded here: https://github.com/danielknox/Robot_Warrior.

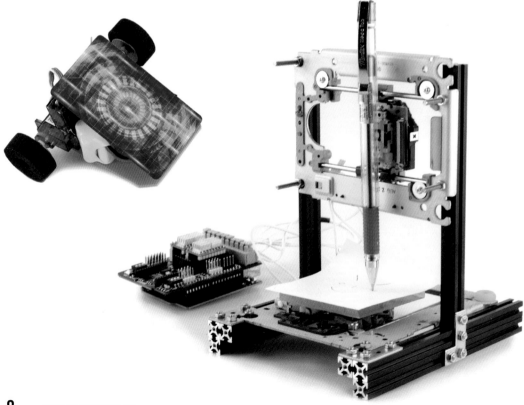

The second project, CNC Writer, is a type of CNC (Computer Numerical Control) machine. In industry, robots are commonly used to perform repetitive tasks over and over again with high precision. Finding the parts for a CNC machine might sound really expensive, but everyday objects around the home actually use similar components. So for this project we will 'go green' and recycle some old discarded electronics to build our own mini CNC Writer. This machine functions like and understands the same language (GCode) as its larger counterparts. Like our other advanced projects, I have prepared the software files for you, so you just need to build the physical machine. The files that you will need to go with our step-by-step guide can be found here: https://github.com/danielknox/CNC_Machine.

The real Mars Rover must rank as the ultimate remote-controlled robot. The final project in the book shows how to build your very own machine that can roam about rough terrain. It is controlled from your mobile phone. At its heart, this robot uses a board based on the popular Arduino platform – this board can easily be expanded to accept additional sensors and there are also many popular software libraries available to help add new functionality. Arduino commonly uses a text-based programming language known as 'C++'. This programming is too long and complex to be described in this book, so to keep things easy and fun for you I have already done the hard part!

All the software that the robot needs can be found here: https://github.com/danielknox/Mars_Rover.

TOOLS GUIDE

Some basic tools and a few household items are all you need to start building your own robots. Tools are also useful for doing any running repairs that might be needed.

Marker pen

Cross-head screwdriver

Flat-head screwdriver

Nut driver

Engineer pliers

Precision snips

Retractable knife

Multimeter

Glue gun

Crimping tool

Wire stripper

Safety glasses

Cordless drill

Drill bits

Drilling vice

Rotary multi-tool

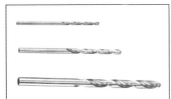

Cut-off disc

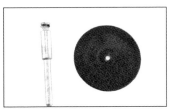

Soldering iron and solder

Helping hands

Hex/Allen keys

Computer

COMPONENTS GUIDE

Every robot featured in the book has a specific list of required components, but here's an overview of some of the most important bits of kit that you will need to acquire.

AAA battery holder with wires

AA battery holder with wires

Switchable AA battery holder with wires

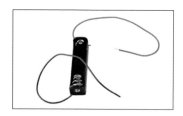

AA 4.5-volt battery holder with switch

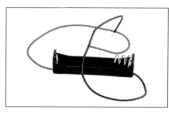

AAA switched battery holder for BBC micro:bit

Modified battery box

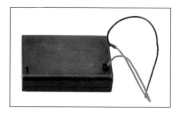

Tamiya battery cable

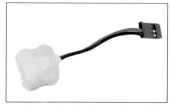

4.8V NiMH battery

NiMH battery charger

20 AWG and 22 AWG stranded wire

Ring electrical crimp terminals

Terminal strip

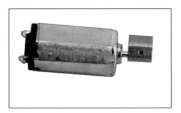

Mobile phone vibrator motor

3-volt motor

6-volt geared robot motors with wheels

SPDT momentary switches with long lever

Continuous rotation servos

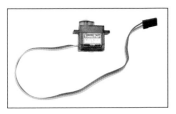

9g servo

Servo tester

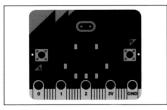

BBC micro:bit

Arduino Uno

DFRobot bipolar stepper controller

DFRobot Romeo BLE

PIR sensor

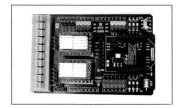

Metal tin (e.g. a mini survival tin)

MakerBeam

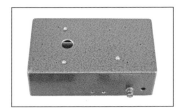

Project box

Bolts, nuts and washers

Screws with nuts (e.g. socket head)

90-degree-angle L brackets

45-degree-angle MakerBeam brackets

5mm threaded rod

Hex standoff spaces

BASIC TECHNIQUES

Step-by-step guidance is provided to show you how to build each project in this book, but there are a few basic techniques that are worth mastering from the very beginning.

SOLDERING

1. Plug in and heat up soldering iron

2. Tin the tip with a thin layer of solder

3. Apply tip of hot iron to solder

ROTARY TOOL

1. Insert cutting disc and tighten

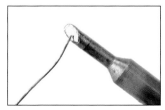
2. Grip item securely in vice

3. Cut part at correct angle

GLUE GUN

1. Plug in and heat up glue gun

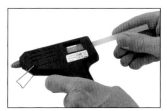
2. Insert glue stick

3. Dispense hot glue

01

BASIC BOTS

BRISTLE BOT

SQUIBBLE BOT

BUTTERFLY BOT

ROBO ROACH

SPIRO BOT

These five fun projects use simple components and everyday objects that you may have around the house. Start building!

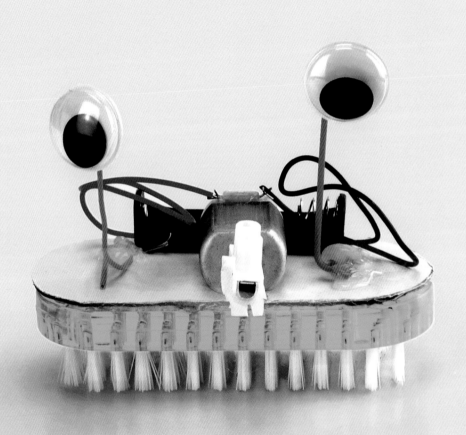

BRISTLE BOT

You don't need many parts or any specialized tools to make Bristle Bots; as a result, they are extremely easy to build and are a great place from which to set out on your journey to become a robot master.

COMPONENTS

1 AAA battery holder with wires
2 paperclips
1 nail brush
1 3-volt motor
2 googly eyes
1 terminal strip
1 piece of cardboard

TOOLS

retractable knife
flat-head screwdriver
marker pen
engineer's pliers
scissors
hot glue gun

SNAPSHOT

Some Bristle Bots can become quite fancy, with steering controls and high-performance brushes taken from electric toothbrushes; however, at their heart, all Bristle Bots are generally built with the same basic parts: a brush and a vibrating motor. So, grab some bits and then let's get started with building our first robot.

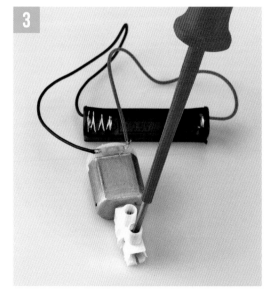

1 Take the AAA battery holder and wrap the exposed wire ends around the 3-volt motor's terminals – they're the bits that look like little brass 'ears' (A). These types of motors don't mind what way around you wire them, so don't worry about which wire goes to each terminal. You can use a soldering iron to secure the wires permanently in place, but twisting the wires around the terminals works too. If you do use a soldering iron, be careful as these tools get very hot.

2 Take the terminal strip and use a knife to cut-off a single segment; be careful as the knife blade is very sharp.

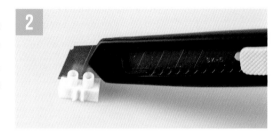

3 Then with a flat-head screwdriver loosen one of the terminal's screws. Slip it over the 3-volt motor's shaft (the bit that turns) and tighten the screw back up. Check to make sure that the shaft can still turn freely; if it doesn't, undo the terminal's screw, ease it away a little bit from the motor's body and then retighten it.

4 To fashion the Bristle Bot's eye stalks, unbend one of the paperclips so that it is roughly straight. Now create a small loop at one end of the paperclip – the loop does not need to be perfect. At the opposite end, form another loop, offset at roughly 90 degrees to the first. To neaten it up, use the cutting edge of a pair of pliers to remove any excess wire. Repeat this step for the second paperclip.

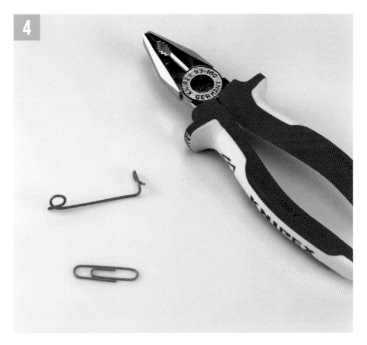

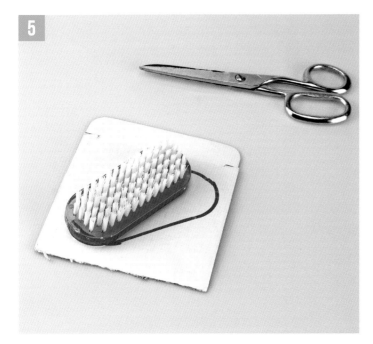

5 Take a sheet of thin cardboard and place the nail brush upside down on top of it. Hold the nail brush securely in place while you trace around it with a thick marker pen. Lift the nailbrush off the cardboard. Now use a pair of scissors to cut around the shape. The cut-out doesn't need to be perfect, but try to make it roughly the size, or slightly smaller than the nailbrush.

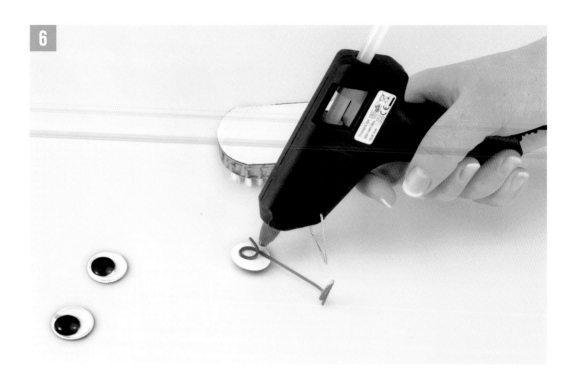

6 You now have several parts that we need to glue together in order to assemble the Bristle Bot. For the next step, it helps to place a large sheet of cardboard on the table that you are working on, to protect it from any hot glue that may drip. Plug in your hot glue gun and wait for it to heat up; be careful as hot glue guns get very hot. When it seems to be ready, take a glue stick and place it in the hole at the end of the glue gun. Pull the trigger and if the glue comes out of the nozzle, you're ready to go. Be very careful as the glue remains very hot for a while after it is dispensed.

Take the cardboard cut-out and using the hot glue gun, deliver a bead of hot glue to cover the edge and middle of the cardboard.

As the hot glue hardens quickly, you'll need to work fast – don't worry about completely covering the cardboard with glue, a little bit is fine! Put down the glue gun and press the sticky side of cardboard onto the nail brush. Choose the side with the fewest amount of bristles for this, otherwise your robot won't have any legs!

Next take one of the paperclips and using the hot glue gun, attach one of the googly eyes to one of the looped ends, repeat this step with the second paperclip. We've nearly finished constructing the Bristle Bot.

7 Now take the motor and battery holder you assembled earlier and glue these to the top of the cardboard cut-out; the terminal segment should hang off one of the long sides of the nailbrush (A). Place and glue the AAA battery holder behind the motor (B). Finally, glue each of the eye stalks to the cardboard, one on each side of the motor.

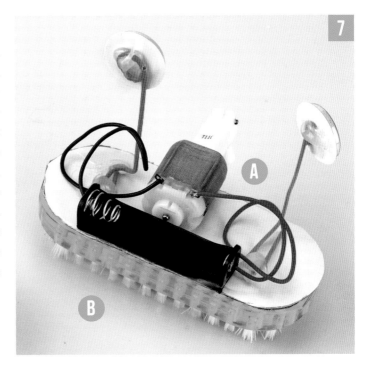

HOW IT WORKS

Insert an AAA battery into the battery holder and the motor should begin spinning. When placed on a smooth surface, like a table-top or tiled floor, the Bristle Bot should shuffle randomly about. Just be careful that it doesn't fall off your table. The bristle moves because the terminal block unbalances the motor's rotation, which in turn causes a lot of vibration to be transmitted through the nailbrushes' bristles. The minimal surface contact of the bristles with the smooth surface means there is little friction, allowing the Bristle Bot to move around easily. Normally motors vibrating in this way could pose a serious problem, but in this case we need this vibration to make our Bristle Bot move.

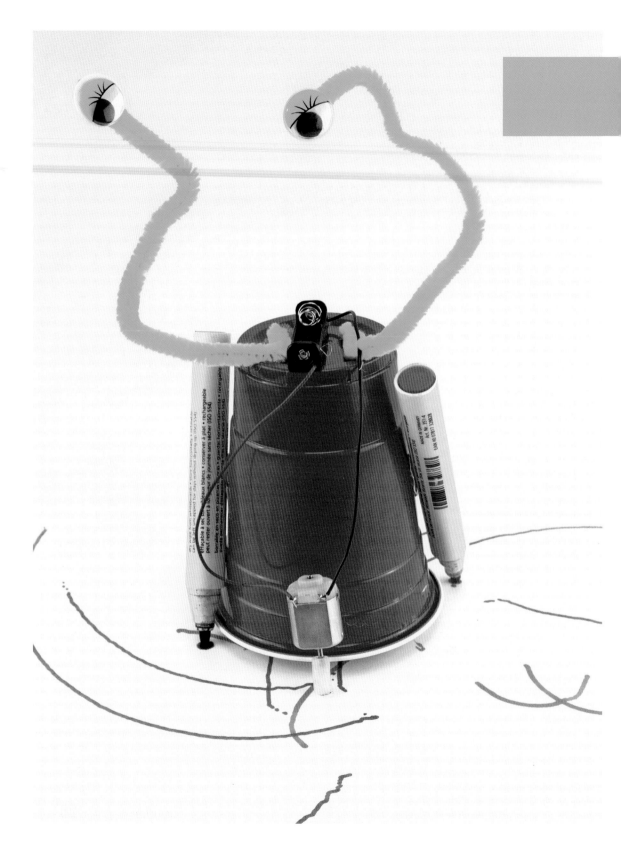

SQUIBBLE BOT

In our last project, we used a vibrating motor to cause our robot to move about on a flat surface. Here we add some extra bits to create a robot that doodles.

COMPONENTS

1 AAA battery holder with wires
1 3-volt motor
2 googly eyes
2 or 3 differently coloured whiteboard markers
1 large plastic party drinks cup
2 colourful pipe cleaners
1 glue stick
1 nail (about the thickness of the motor shaft)
1 large sheet of paper (at least A3)

TOOLS

retractable knife
cutting mat
hot glue gun
small piece of sandpaper

SNAPSHOT

It is a nice idea to attach pens to bristle bots, so that they make all sorts of colourful doodles while they jiggle about.

For this project, instead of using a vibrating motor, we will craft a small 'wheel' to create a robot that spins around – this gives our robot the ability to make snazzier doodles.

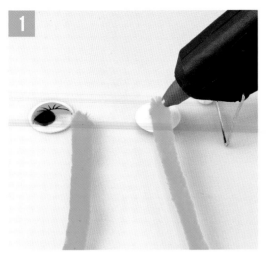

1 Plug in your hot glue gun and wait for it to heat up. While it is warming up, bend each pipe cleaner at one end so that the bend is roughly 2.5cm (1in) long.

When the glue gun is ready, insert a glue stick into it. Dispense a small bead of hot glue onto the back of each googly eye and then attach the long end of each pipe cleaner to it. You now have two eye stalks for your robot.

BRAINWAVE

Protect your work surface by placing a large sheet of cardboard on the table that you are working on; this will help to shield it from any hot glue that may drip..

2 While the glue gun is still warm, dispense a small amount of glue to the short end of each eye stalk and attach them to the top of the plastic drinks cup; you need to leave about a 2.5cm (1in) gap between them. When the glue has hardened, bend your eye stalks to give your robot a more 'alien' look. You can now unplug the hot glue gun, but don't put it away as we will need it again later.

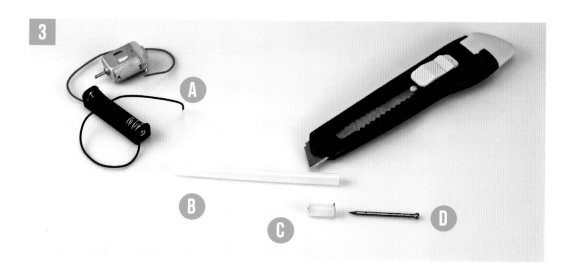

3 We are now going to assemble the motor for our robot (A) to let it walk about. First take a spare hot glue stick (B) and use a retractable knife to cut a short bit off (around 1.2 cm or ½ inch); make sure you do this on a cutting mat, so you don't damage the table. We want to slip this cut-off (C) onto the shaft of our 3-volt motor, but before we can do that we need to make a small hole in one of the ends. The easiest way to do this is to slowly twist a small nail (D) into the hot glue stick – if it's too hard to go in, try gently heating the glue stick a little to soften it.

Once you have made a small hole, slip the cut-off portion onto the motor's shaft. It doesn't have to fit all the way onto the shaft, but you want it gripped tightly. You can test this by trying to spin the glue stick and making sure it doesn't come off the shaft.

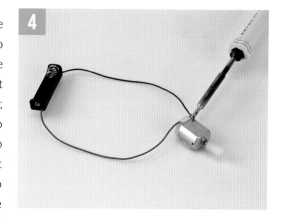

4 Take the AAA battery holder and wrap the exposed wire ends around the 3-volt motor's terminals. As with our previous robot, it doesn't matter which way around the wires go to the terminals. You may wish to use a soldering iron to secure the wires permanently in place, but twisting the wires around the terminals works well too.

5 We now have all the parts we need to assemble our robot, so switch the hot glue gun on again and get some spare sticks to hand – we are about to do a fair bit of glueing.

First glue the AAA battery holder onto the top of the plastic drinks cup; it should fit neatly between the eye stalks that we stuck on earlier.

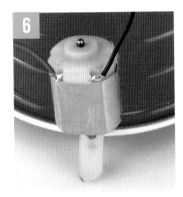

6 Next, glue the motor onto the bottom of the cup. You want the hot glue stick cut-off that's on the motor's shaft to rotate freely below the bottom of the cup.

It helps to use a small piece of sandpaper to roughen up one side of the motor – apply hot glue to the body of the motor and then press it onto the drinks cup.

HOW IT WORKS

First place a large sheet of paper on a smooth surface, such as a tiled kitchen floor. Hold the robot in the air and insert an AAA battery into the battery holder – the shaft of the motor should begin spinning. When placed on the centre of the sheet of paper the robot should begin to spin and skitter about; make sure you're ready to catch it before it runs off the paper, as you don't want the robot to mark the table or floor.

The robot moves because the motor is rotating counter to the body of the drinks cup, and this causes the cup to begin to rotate. The glue stick acts like a small wheel, its surface helps to provide grip between the motor and the surface of the paper – it also flexes slightly on the shaft allowing the robot to skitter around.

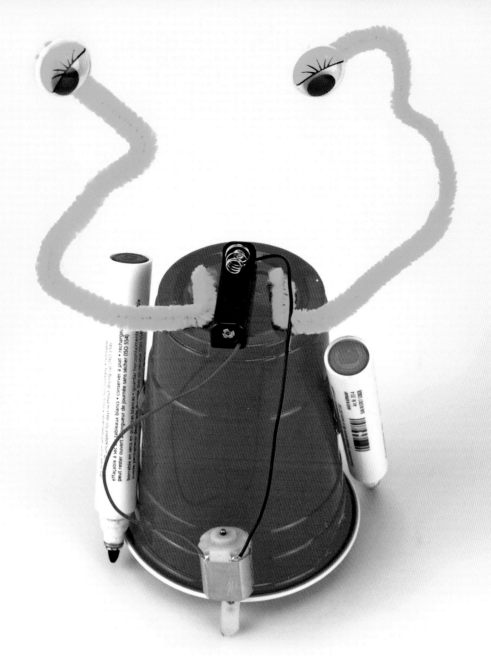

7   Finally take the caps off two or three whiteboard markers and dispense a bead of glue along one side of the body of each pen. Attach them to the drinks cup at a 45-degree angle; the tip of each of the pens should come down to roughly the bottom of our cut-off piece of glue stick. Space the pens evenly around the body of the drink cup.

BUTTERFLY BOT

Our first Bristle Bot was cute, but it's currently rather lonely. This next creation is mechanically similar to the first one, but it uses a much smaller motor to produce the required vibrations.

COMPONENTS

1 AAA battery holder with wires
2 toothbrushes (toothbrushes with angled bristles work even better than straight ones)
2 googly eyes
1 mobile phone vibrator motor
1 colourful craft bead
spool of enamelled craft wire
sheet of 'fantasy' film
PVA glue
lead-free solder

TOOLS

drill vice
safety glasses
engineer's pliers
wire strippers
scissors
helping hands
hot glue gun
soldering iron

rotary multi-tool
  (e.g. Dremel)
tealight (fairy) candle
lighter/matches
brush
small piece of
  sandpaper

SNAPSHOT

This bristle bot is a great introduction to some important tools, particularly the rotary multi-tool and the soldering iron.

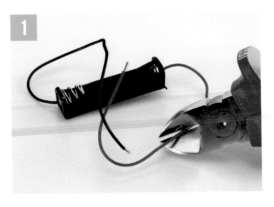

1

Plug in your soldering iron – be very careful as soldering irons get very hot (they need to be able to melt a soft metal). While it is warming up, use the pliers to cut the wires of the AAA battery holder to a shorter length, leaving one wire longer than the other. Strip the insulation off the ends of the wire to expose a small amount of bare wire.

2

Once your soldering iron has reached the right temperature it's time to connect the two wires to each of the terminals on the small phone vibration motor. First 'tin' the exposed wire ends; use the helping hands to hold the two wires and then use the soldering iron to melt the solder onto the exposed wires. Remove the wires from the helping hands.

Grip the vibration motor in one of the crocodile clips of the helping hands, so that you are able to work easily on the two small metal terminals. Touch one of the terminals with the tip of your soldering iron and gently press one of your tin wire ends against it (A). When the solder melts, move your iron away to let the solder cool; once it does, you can stop pressing the wire against the terminal. Repeat for the second wire.

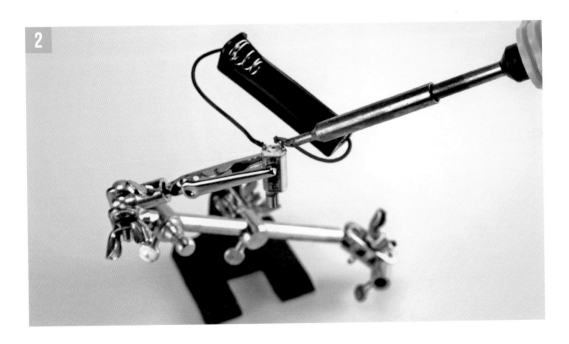

3 Put your battery and motor to one side. Grip a toothbrush in the drill vice so that the bristles overhang the end of the vice. Insert a cut-off disc into the chuck of your rotary tool. Put on your safety specs and then plug the rotary tool into an electrical outlet and set it to a high speed. Cut off the end of the toothbrush using the rotary cutter. (You don't need to press down hard on the toothbrush to make the cut, the tool will do the work for you.) Repeat for the second toothbrush.

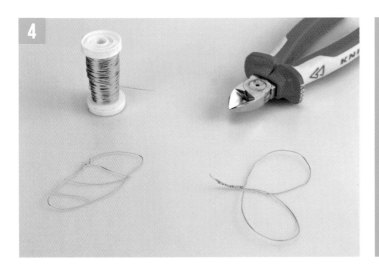

4 Put the toothbrushes to one side and unplug your rotary tool. We are now going to make two wings for our bristle bot – this same process can be used to make wings for all sorts of toys.

Unwind a length of wire from the spool of enamelled craft wire (around 15cm/6in) and cut it using the pliers. Form a loop with the enamelled wire, and twist the ends to stop it unravelling (B). Cut another length of wire and form a second smaller loop. Twist the ends of both loops together, and trim the ends with your pliers if that's needed.

5 Use a brush to apply a small amount of PVA glue to the wire (not too much) and stick it down on a small sheet of fantasy film – you may need to press down gently on the wire to make sure it sticks to the film. Be careful not to move the wire around too much as the glue will smear on the surface. Let the glue dry (this will take a few hours). Once it has dried, use a pair of scissors to cut around the wings – leave a bit of excess film on the edges because it will shrink and harden when it is heated. Repeat steps 4 and 5 to make a second wing.

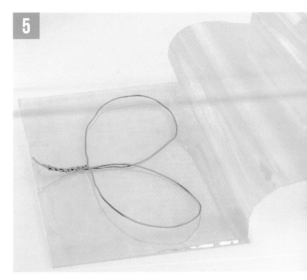

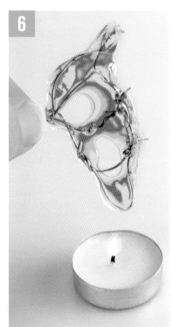

6 Use a match to light a tealight candle. Gently bring your fairy wing near to the flame (but not actually touching it). The heat of the flame will cause the film to harden and shrink. You will notice that the film goes through a range of brilliant colours, but be careful – if it gets too hot, holes will form. To avoid this, gently move the wing above the flame and don't spend too long on this step (a couple of seconds is more than enough). Repeat this step for the second wing.

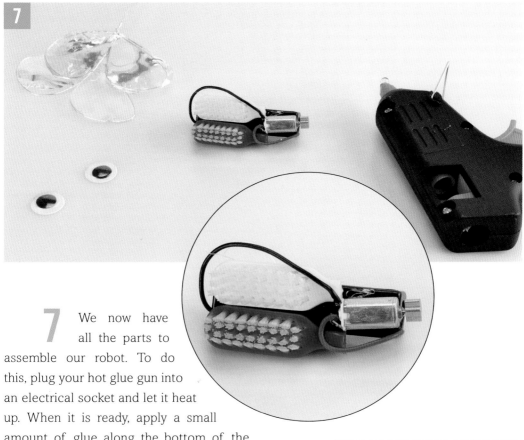

7 We now have all the parts to assemble our robot. To do this, plug your hot glue gun into an electrical socket and let it heat up. When it is ready, apply a small amount of glue along the bottom of the AAA battery holder. Attach the two toothbrush ends to one end of the battery holder. Now use a small piece of sandpaper to roughen up the flat surface of the small mobile phone vibrator (this will help it to stick). Dispense a small amount of glue onto this surface and attach the motor to the bottom of the AAA battery holder (it should sit behind the two toothbrush ends). Flip the robot over and attach the two wings using a small bead of hot glue; you will have to support them while the glue is cooling.

Finally, apply a small amount of glue to each of the two googly eyes and attach them to the front of the robot (away from the motor end!). A small bead can also be added to give the Butterfly Bot a cute nose!

HOW IT WORKS

Insert an AAA battery into the battery holder and the small motor should begin spinning. When placed on a smooth surface, such as a table-top or tiled floor, the Butterfly Bot will start to shuffle randomly about – just be careful that it doesn't fall off your table.

ROBO ROACH

Many robots use small computers known as microcontrollers to sense and adapt to their environment. However, it's also possible to make a reactive robot that only uses mechanical parts (so you don't need to write any computer code).

COMPONENTS

1 AA battery holder
1 magician robot chassis kit
2 6-volt geared robot motors
   with wheels
2 SPDT momentary switches
   with long lever
2 paperclips
length of 20 AWG stranded wire

TOOLS

soldering iron with stand
solder (lead-free)
hot glue gun
cross-head screwdriver
wire strippers
helping hands
retractable knife

SNAPSHOT

Our robot cockroach is exactly that – it's able to explore its surroundings and move around any obstacles that it encounters just by using two switches and some wires. Building a robot chassis can be quite difficult for a beginner roboticist, so we will make our lives easier by adapting a simple off-the-shelf chassis that is easy to come by. However, if you're feeling adventurous, it possible to build a similar chassis using some thick cardboard and the liberal use of adhesive from a hot glue gun.

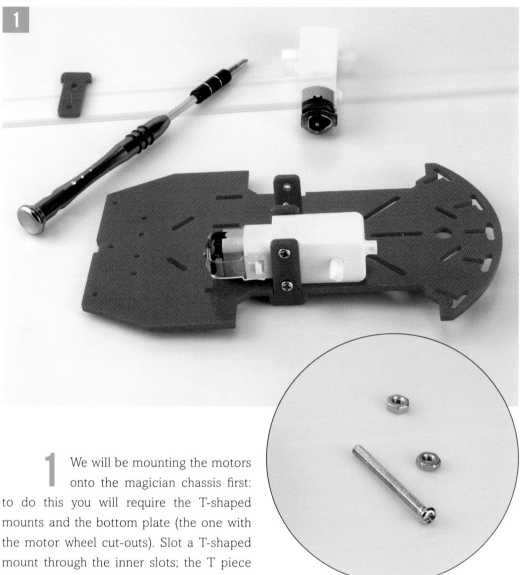

1 We will be mounting the motors onto the magician chassis first: to do this you will require the T-shaped mounts and the bottom plate (the one with the motor wheel cut-outs). Slot a T-shaped mount through the inner slots; the T piece should prevent the mount going all the way through the bottom plate.

Align one of the 6-volt motors against the mount that you have just installed – the metal motor of the motorized gearbox should face towards the straight-edged rear of the bottom chassis. Take another T-shaped mount and place this against the motor. Push a bolt through the mounting holes and screw on a nut on the opposite side – this can be fiddly, so insert the bolts and loosely attach the nuts before tightening everything.

Repeat this step for the second motor.

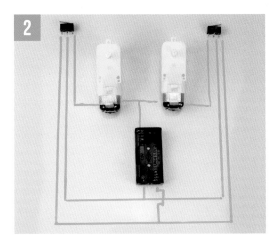

2 We are now going to wire the motors to the battery and the switches. The next few steps take some time and are a little complicated, so it is worth studying the picture on the left to understand where the wires will be going.

Plug in your soldering iron and allow it to heat up.

3 Cut a length of wire, so that it is long enough to run between the two motors and expose a portion of wire using a pair of wire strippers. Loop the wire strands around one of the terminals on each of the two motors.

Cut and strip another length of wire, so that it is will reach from one of the motors to the front of the plate (around 10cm/4in). Loop its wires through one of the terminals that already has wire strands wrapped through it and then thread the other end through a nearby hole in the chassis to emerge on the front side of the plastic plate. Solder all of the wires to their terminals, so that they are secure.

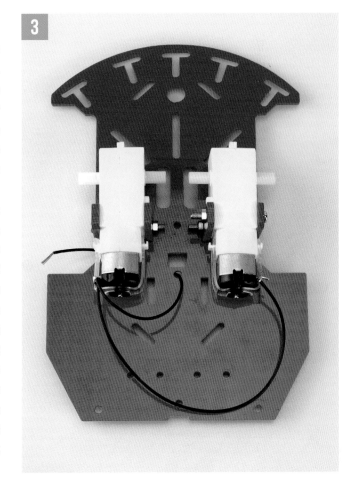

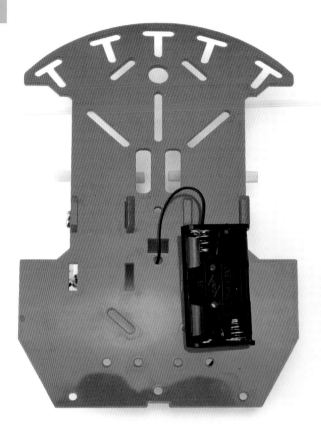

4 Flip the plastic plate over and hold your AA battery holder against it. Wrap the loose wire end around one of the positive terminals – this type of terminal is the one that looks like a spring. If you wish you can solder this wire in place, but be careful not to melt the plastic. Once you have done this, put the plastic plate to one side for the time being.

5 We will now solder wires to each of the switches. Put one of the switches into a pair of helping hands, and melt a little bit of solder onto each of its legs – this will make it much easier to attach the wires in a minute. Now cut and strip three lengths of wires, each approximately 15cm (6in) long. Put one of the wire strips into the spare croc clip on the helping hands and melt a little bit of solder onto the exposed wire. Repeat this for each of the wires.

Finally solder the 'tinned' end of each wire onto one of the legs of the switch (A), make sure you don't 'bridge' any of the connections (each wire and solder joint should only touch one leg).

Repeat this step for the second switch.

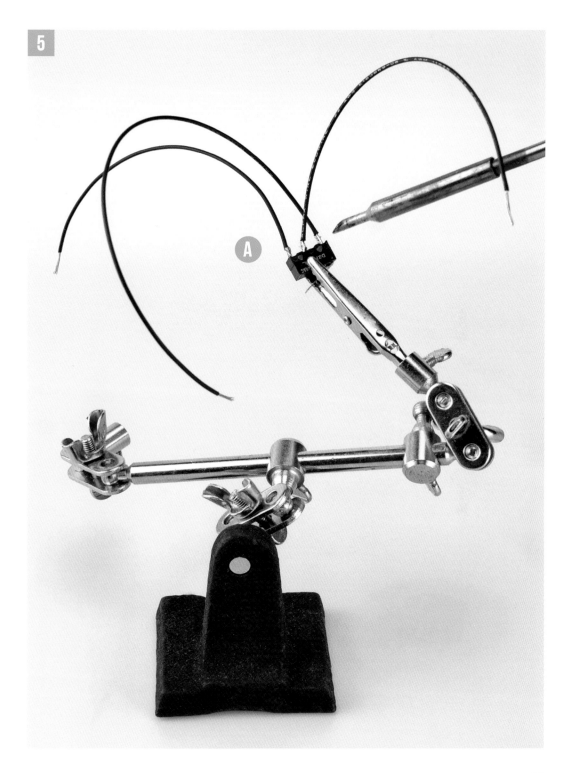

6 Look at each switch closely and you should see that each leg has code letters above it: NC (Normally Closed), NO (Normally Open) and C (Connected). The reason for these three terminals is that terminal 'C' changes what it is connected to, depending if the switch is being pressed down or not.

Take the wire running from one of the terminals labelled 'C' and use the wire strippers to strip the non-exposed end of the wire. Loop this end around one of the non-soldered motor terminals. Repeat this step for the other switch, looping its wire around the remaining motor terminal. Solder each of these terminals.

7 Use the wire strippers to expose the wires that connect to the 'NC' terminal on both of switches. Twist the two exposed wires together and use the soldering iron to solder the twisted pair of wires.

Flip the plastic chassis plate over and find the terminal that is opposite to the one that we soldered earlier (a non-sprung terminal). Melt a small amount of solder onto this terminal, but be careful not to metal the plastic. Now solder the twisted pair of wires to this.

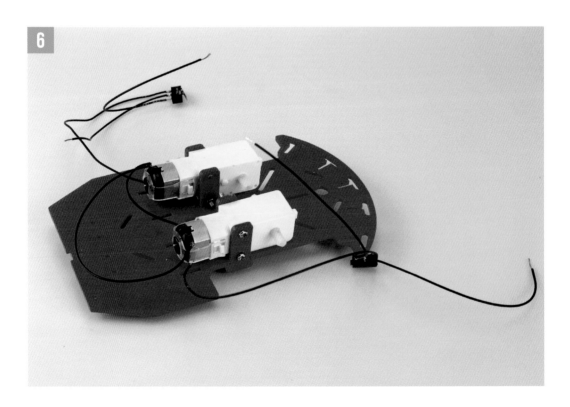

6

7

8 The remaining wires from the terminal marked 'NO' (A) should now be stripped and twisted together.

8

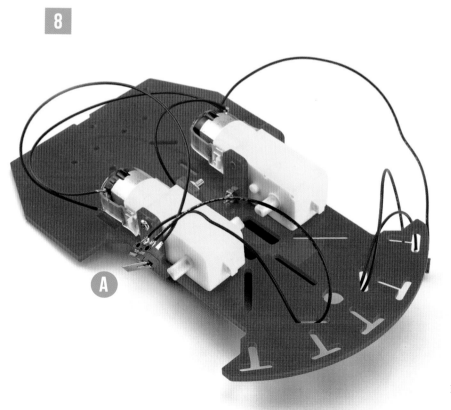

A

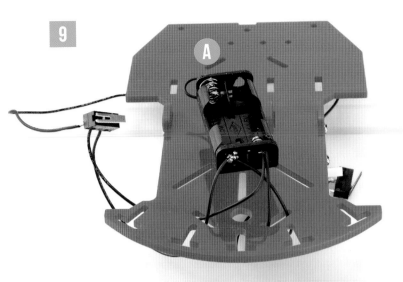

9 Now solder this pair of wires onto the remaining positive terminal (the spring-like one) of the AA battery holder (A).

That's it! We have now completed all of the soldering. Time to unplug and put away the soldering iron.

10 Plug in the hot glue gun and let it warm up. Flip the plastic plate back onto its underside so that you can see the motors. Gently bend the wires so that each of the switches reaches the curved edge of the motor plate at the front of the bot. The highest point of each of the metal levers should point towards the side of the plastic plate that is nearest to them.

Once the glue gun is hot, glue each of the two switches to the plastic plate.

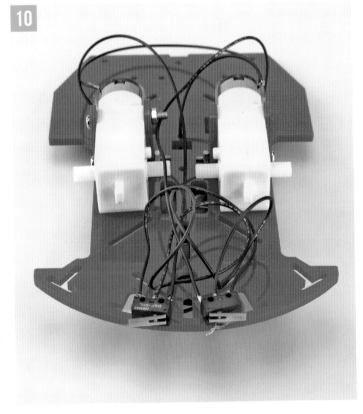

11 Flip the plastic plate over and take the risers from the magician chassis kit – they look like long hexagonal pieces. Attach these to the plastic plate, one in the middle of the chassis and one on each side – you should be able to imagine drawing a triangle between these risers.

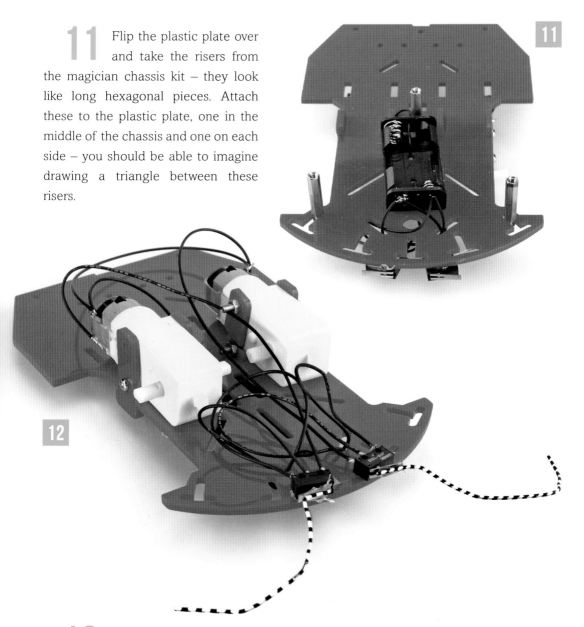

12 Unbend each of the paperclips and form them so that they are shaped like insect feelers. Apply some hot glue to each of the switch levers and attach a 'feeler' to each one. Press down to make sure the feelers are firmly attached.

Once the glue has hardened, make sure that the switch is still able to toggle on and off; if it doesn't, use a knife to break away some of the hardened glue so that it moves more freely.

13 Take the wheels from the magician chassis kit and slip them onto the plastic mounts of each of the motors. You should also have a caster ball wheel in the kit (A); install this onto the back of robot chassis (the end that is opposite the switches).

14 The remaining plastic plate can now be attached to the metal risers. Flip the robot over and place the plastic plate on top of the risers. Adjust the plate so that it is aligned with the bottom plate and then screw it into place. After all that hard work, our robot is complete!

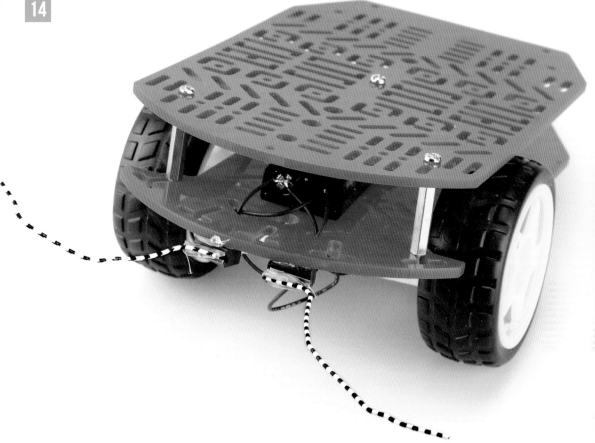

14

HOW IT WORKS

Remove the top plastic plate to insert some batteries into the battery holder – keep a firm hold of the Robo Roach so that it doesn't run away. Screw the top plate back onto the robot. Put the roach onto a smooth floor – if it starts to drive backwards you may have to swap the motors around! When the roach encounters an obstacle, one of the motors should stop turning because the paperclip feeler will move and so flip the switch temporarily to an off position. This causes the roach to turn. Much like larger, more expensive robots, our roach's sensor can only 'see' certain types of obstacles; it should be fine when reacting to walls and objects with large surfaces, but it might get tripped up by smaller thinner items, such as table legs.

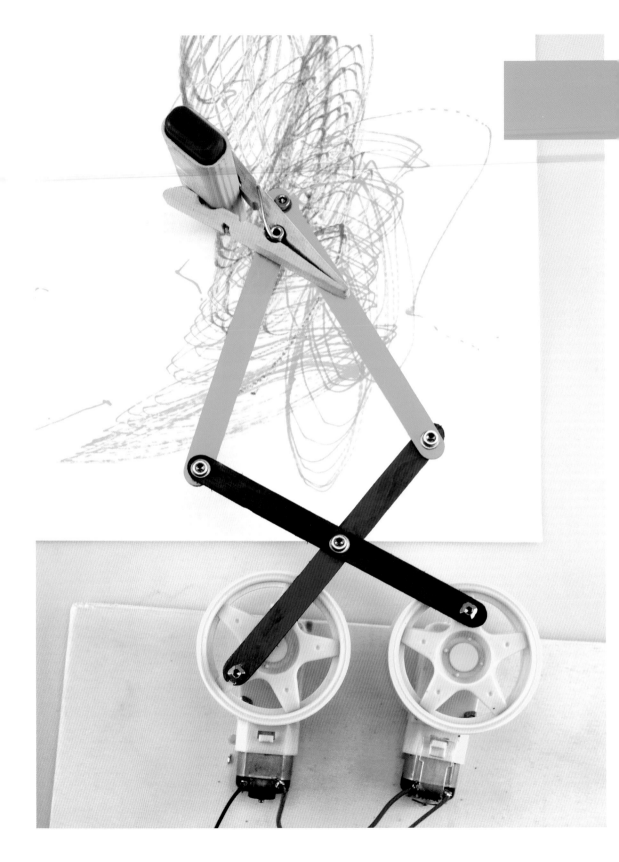

SPIRO BOT

Our earlier Squibble Bot produced marks on paper, but the results were fairly random. The robot that we are about to build has much better drawing skills than our Squibble Bot and is able to replicate the results obtained from a classic toy known as a 'Spirograph'.

COMPONENTS

1 switchable AA battery holder with wires
2 6-volt geared robot motors with wheels
1 clothes peg
6 coloured lollipop sticks
1 6mm sheet of wood
6 3mm washers

10 3mm nuts
4 3mm bolts (12mm length)
6 3mm bolts (40mm length)
6 3mm spring washers
length of 20 AWG stranded wire

TOOLS

soldering iron with stand
hot glue gun
rotary multi-tool (e.g. Dremel) with 1-3mm drill bits
drill vice
safety glasses
marker pen
masking tape
flat-head screwdriver
cross-head screwdriver
engineer's pliers

SNAPSHOT

Spirographs produce curves by using mechanical gears, but gears can be hard to fashion without some fancy equipment and some complicated maths. Instead, we will use lollipop sticks and motors to make a simple robotic arm that can move a pen on its own.

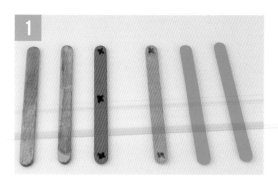

1 Take a lollipop stick and using a marker pen draw a small 'X' in the middle of it. Make two other small marks at each end – you should leave enough room for the 3mm drill bit to be able to make a hole at the centre point of each mark. Take a second lollipop stick and mark a small 'X' at each end (no need for a mark in the middle of this one).

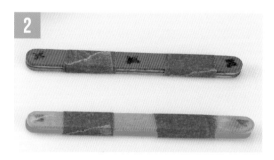

2 Place the first lollipop stick on top of two unmarked sticks and wrap some masking tape around them to secure them tightly together; keep the tape clear of the marks as you need to be able see these. Do the same for the other marked lollipop stick. You should now have two neat bundles of sticks.

3 We are now going to drill holes into each of the lollipop stick bundles, using the marks as guides. First place one of the lollipop stick bundles into a drill vice, so that it is held securely in place; you don't want to drill into the vice, so let it overhang a little bit.

Set up the rotary multi-tool with a 1mm drill bit; to do this, loosen the tool end of the rotary tool, so that you can insert the drill bit into the collet (the gizmo that holds the tool in place), then retighten the nut. If the drill bit doesn't fit, you'll probably have to swap the collet for a different sized one.

Put on your safety specs and plug the rotary tool into an electrical socket. Set the rotary to a relatively fast speed and drill into the wooden lollipop stick at one of the marks – don't force down hard on the drill head as these tools are not designed for that kind of use; just gently press down on the lollipop stick.

When you are through the bundle of sticks, stop the tool and repeat for the other marked holes. The holes you have just made are called 'pilot' holes – they help guide larger drill bits. Once you have drilled the first set of holes, replace the drill bit with a 3mm sized one. Repeat the process again, drilling each hole out neatly.

4 Put the bundle of lollipop sticks to one side and remove one of the robot wheels from the 6-volt geared motor. Place this wheel into your vice and tighten it so that the wheel is secure. You will notice that the wheel hub has several dimples, one on each spoke. With your rotary tool, drill a hole through one of these dimples; as before, make a pilot hole first, before you use the 3mm drill bit. Repeat this with the second 6-volt motor wheel.

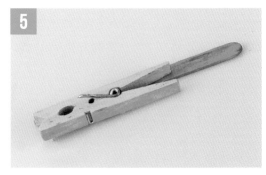

5 Plug in your hot glue gun; remember to protect your work surface. While the glue gun is heating up, unwrap the bundles of lollipop sticks.

Take one of sticks that has only a hole at each end and put a three-quarter-length bead of hot glue along one of the edges of the lollipop stick. Stick the clothes peg onto this bead of glue, so that it overhangs the stick. When the glue hardens, you should still be able to open and close the clothes peg normally. Once this is done, you can unplug the glue gun.

6 We are now going to make the 'scissor' mechanism for our Spiro Bot. Take a 3mm bolt and put a washer on it, then put this bolt through our clothes peg lollipop stick; start with the hole nearest the clothes peg. Take another lollipop stick that only has two holes and put the bolt through one of its holes. Now add a spring washer onto the bolt and finally a nut. Tighten the bolt. The washer allows the mechanism to move smoothly and the spring washer prevents the nut from loosening. All of our holes will observe this same sequence: bolt, washer, lollipop stick 1 and 2, spring washer and a nut.

Take the two lollipop sticks that have three holes, and make a 'X' shape with them – the middle holes should line up with each other. Fasten these together with a bolt, washer, spring washer and a nut. Now attach this cross shape to the clothes peg mechanism that we just made; overlap the holes and then put a bolt, washer, spring washer, nut combination through each of them. We have nearly finished this step.

Finally take the two wheels that we drilled holes into earlier and using a longer bolt attach each wheel to the remaining holes on our lollipop stick scissor mechanism. Tighten all of the bolts using a screwdriver.

We have now finished assembling the scissor mechanism – nice one!

7 Put the two 6-volt motors onto the edge of the sheet of wood about 10cm (4in) apart from one another. With a marker pen, mark around the edges of the motors.

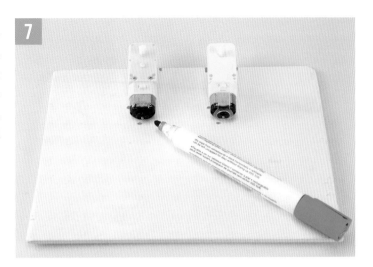

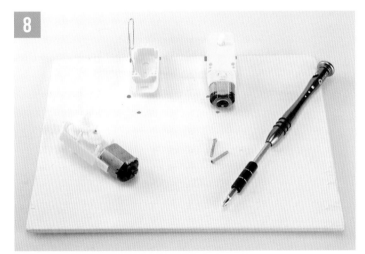

8 Take apart one of the motors with a cross-head screwdriver. Position the shell of the motor using the markings on the wood as a guide. With a scribe (or a paperclip), mark the position of the motor mounting holes onto the wood plate – do this for both motors.

9 Move the motors out of the way and get your rotary multi-tool out again. Using the rotary tool, drill each of the mounting holes that you marked on the wooden board; don't forget to protect your work surface from the drill and to wear your safety glasses.

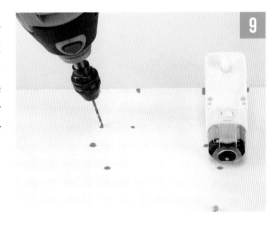

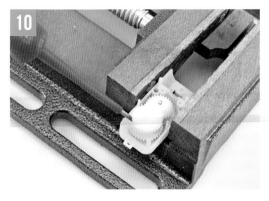

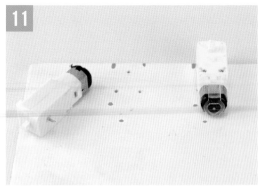

10 You will notice that the motors do not lie flat on the wooden surface because there is a white wheel mount coming out of both sides. Although each side of the motor looks very similar, they are actually subtly different – one has a small raised dome and the other side doesn't. On the side without a raised dome it is possible to pull the white mount off. Do this to each of the motors. I find it helpful to take each motor apart and gently use a screwdriver to push up on the internal gears – the white wheel mount should fall off.

11 Reassemble each motor and put them back on the wooden plate. Line the holes that you made in the board up with the mounting holes of each motor, and then put a 40mm bolt through each one and secure it in place with nuts underneath the board. Tighten these up, so that the motors are held securely to the wooden board.

HOW IT WORKS

Put a sheet of A4 paper in front of the robotic arm, and make sure that the tip of the pen still touches the paper when the arm is fully retracted (you can manually turn the wheels to do this). Secure the piece of paper to your work surface using a little bit of tape on each of the corners. Put two fresh batteries into the AA battery holder and move the switch to its 'On' position. The arm will now begin moving and producing a drawing. When you're happy, switch off the Spiro Bot. If you wish, you can move it to a new position on the paper and start it again – this can produce some really cool-looking patterns.

12 Plug in your soldering iron and let it warm up. While it is doing so, cut a short length of wire with the cutting edge of the engineer's pliers – the length should be enough to stretch between the two motors. Loop the wire through one of the terminals on each of the motors and solder it in place – if the motor terminals are facing towards the wooden panel, it is possible to remove the motor from the gearbox and swap it round. Now take the AA battery box and solder one lead to each of the remaining motor terminals. You can now switch off the soldering iron.

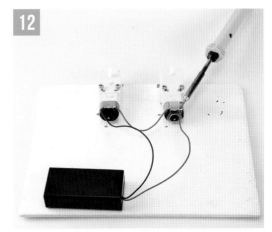

13 Take the scissor mechanism that you built earlier and slot the wheels onto the mounting posts of each motor, the mechanism should hang over the wooden board. Carefully open the jaws of the wooden peg so that its able to grip a marker pen, adjust the pen's position so that the tip of the pen sits just slightly touches your work surface. Job done!

SIMPLE ROBOTS

AVATAR

SCUTTLE BOT

GARDEN GUARDIAN

CATAPULT BOT

WALKING ROBOT

Using the BBC micro:bit to provide the brains for these cool projects, you will soon learn how to adapt the ideas to dream up your own brilliant robot designs.

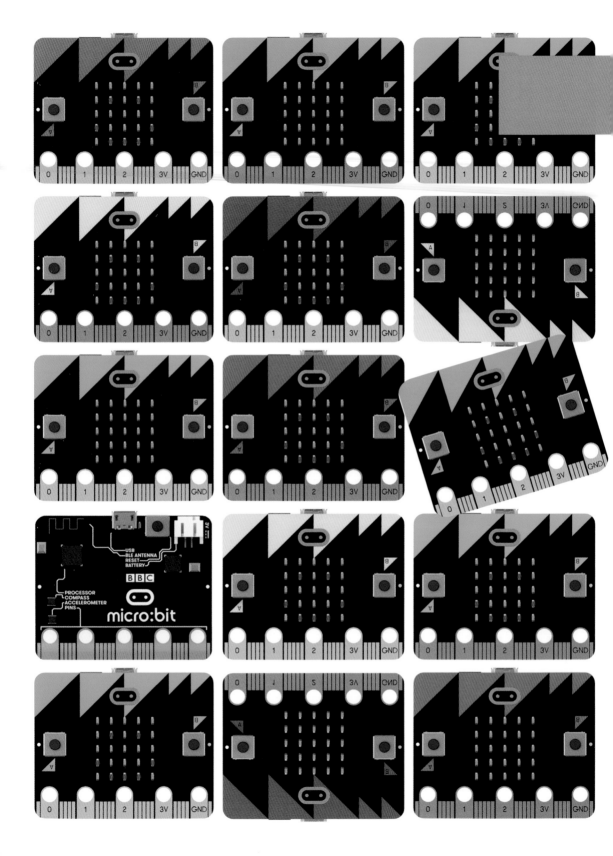

AVATAR

Our previous projects all used simple mechanical and electrical components to allow our robots to move and interact with their environments. However, more complicated robots commonly use small devices known as microcontrollers.

COMPONENTS

1 BBC micro:bit

TOOLS

PC/laptop computer
micro USB cable

SNAPSHOT

Microcontrollers give robots more advanced functionality. They are like small mini computers. While they're not very good for familiar tasks like browsing the internet, they are excellent at doing or monitoring things repeatedly – they are brilliant at this because they can perform these tasks extremely quickly.

Microcontrollers need to be programmed for them to be able to know what tasks they should perform. Programming is commonly done using something that looks a bit like a strange text language, but this is not very user-friendly for new roboticists. Instead we will use a graphical programming language where we can build programs by dragging and dropping 'blocks' onto a screen.

If you have ever used the programming language 'Scratch' at home or at school, you should feel confident about the programming part of these projects. If not, don't worry as this project is all about guiding you through writing your first program.

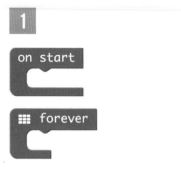

1 Launch the JavaScript Blocks Editor from the micro:bit website (microbit.org). When it loads, you should see a pane with a picture of a micro:bit, a number of tabs that contain different kinds of blocks, and a pane which contains two blocks 'on start' and 'forever'. The 'on start' block, like its name suggests, performs tasks that should happen when your micro:bit is first switched on – this block is useful for setting the initial state of things. The 'forever' block repeatedly executes the same tasks over and over again as fast as the microcontroller can go. This is why microcontrollers are great at doing simple tasks repetitively.

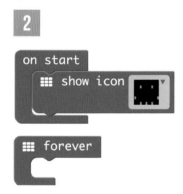

2 Click on the 'Basic' blocks tab and drag a 'show icon' block onto the pane with the 'on start' and 'forever' blocks. You will notice that the 'show icon' block has a little notch in the top part of it, this means that it needs to be connected to something in order to work. Drag and release the 'show icon' block so that the notch connects to 'on start' block. If you have speakers plugged into your computer, you should hear a 'click' sound. The pane with a picture of a micro:bit should now change to show an image of a heart. This pane simulates the running of your program and is useful to test things without having to keep downloading your program to your micro:bit. Click the heart icon that's currently on display in the 'show icon' block and click on the happy face. You should see that the simulator now updates itself and instead of showing a heart, it is instead displaying a happy face.

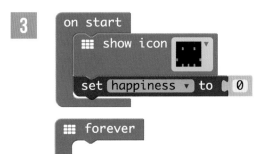

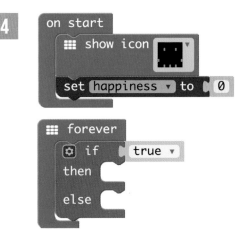

3 Frequently when programming we need our program to remember things; for example, if we wish to count the number of times someone presses a button, the program will need a place to store that counter. This is known as a variable and is one of the most important concepts in many programming languages.

The JavaScript Blocks Editor also has the ability to create variables, so let's make one. First click on the tab labelled 'Variables'. A few blocks should appear, but they are for an existing variable called 'item'. This will not do. Instead, click the button labelled 'make a variable'. When it asks you to give it a name, type 'happiness'. We should now see that the 'Variables' tab contains a new block called 'happiness'. Drag a 'set item to' block to the main pane and connect it to our 'show icon' block. Click on the dropdown menu labelled 'item' and choose 'happiness'. Now when our program first runs, behind the scenes our happiness variable gets set to 0. When programming it is very good practice to initially set your variables to a known value. Something my students often forget to do!

4 We now want our program to do something based on whether or not our happiness variable is less than '0'. In programming this is known as an 'if else' statement. If something is 'true', it will perform an action, if not, it will perform a different action. To do this, click on the 'Logic' tab and drag in the 'if true, then, else' block. Connect this block to the existing 'forever' block.

5 Currently our 'if true, then, else' block is incomplete. This is because it doesn't know what to check to see if the statement is true. We want our program to check whether our happiness variable is less than '0', which means we wish to compare the value of happiness to 0. Click the 'Logic' tab again and drag in the block '0 < 0'. It should connect to the 'if' part of our 'if true, then, else' block.

6 This block is currently checking to see if '0' is less than '0' (something that is false). Click on the 'Variables' tab and drag a 'happiness' block to the screen. Replace the first of the zeros, so that the statement now reads 'if happiness < 0'. Now whenever our happiness variable is less than '0' it will perform one set of actions, and whenever it is equal to or greater than '0' it will perform a different set of actions. Because these blocks are connected to the 'forever' block, this check and set of actions will be performed over and over again.

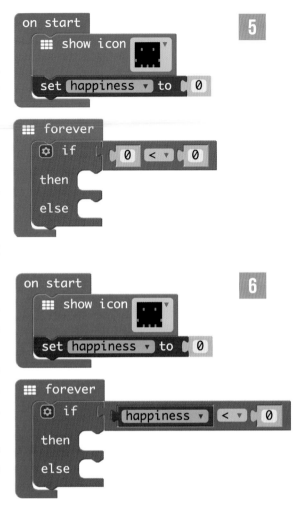

7 Although we are now able to check the value of our happiness variable, you might notice that currently it is only ever changed in one place on start. It is not much good if our happiness is only ever '0', so let's fix that. Click on the 'Input' tab and drag the 'on button A pressed' block to the screen. Unlike the previous blocks that we added, this one doesn't have a notch in its top – this is because it performs actions as soon as something external happens to our micro:bit, in this case, if button A gets pressed. Drag in a second 'on button A pressed' block. You will notice that this new block has been greyed out. This is because we already have a 'on button A pressed' block, so the language is confused why you need a second. Click on the 'A' in this new block and choose 'B' from the dropdown menu – you should find that the block is no longer greyed out.

7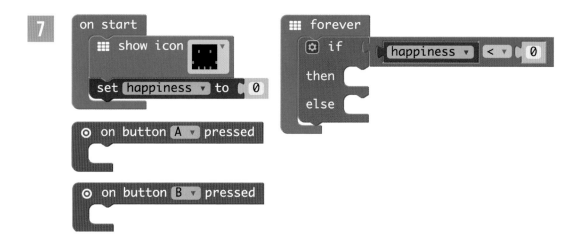

8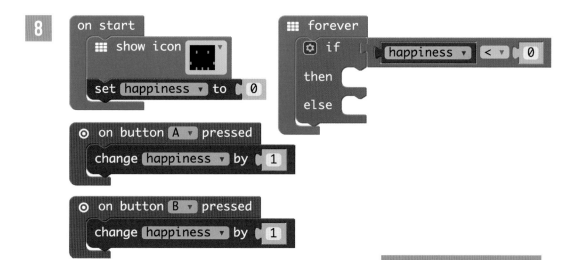

8 Our program now knows that we want to do something whenever we press button A or button B, but it doesn't know what. Click on the 'Variables' tab again and choose the 'change item by 1' block. Drag and connect it beneath our 'on button A pressed' block. As before we are not interested in changing the value of 'item'. So, click on 'item' and instead choose 'happiness' from the dropdown box. Create another one of these blocks and connect it beneath 'on button B pressed'.

BRAINWAVE

Don't confuse the A and B buttons with your keyboard letters. You need to click on the A and B buttons on the micro:bit simulator screen.

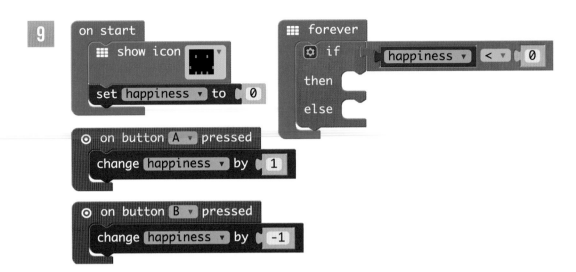

9   The smart roboticist will spot that currently button A and B do the same thing. It will increase the value of the variable 'happiness' by '1'. While a happy robot is always a good thing, it is pointless having two buttons to do the same thing, so change one of the 'change item by 1' blocks, so that it reads 'change item by -1'. Now this button will decrease the value of the variable 'happiness' by '1' whenever it is pressed.

10   Our program is nearly complete. We have a variable called 'happiness' that is initially set to 0 when the program starts, and this can be changed by pressing either button A or button B. We also have a happy face on our micro:bit's LED screen.

To finish our program, let's make something happen whenever our happiness variable is less than '0'. To do this, drag a 'show icon' block from the 'Basic' tab and connect it to the 'then' part of the 'if, then, else' block. Choose an icon that you would like the micro:bit to display. I went with a 'sad' face. Try the simulator out and you should find that the happy face changes when you press the button that was associated with the 'change item by -1' block; however, it currently doesn't change back when we press the other button two times. To fix this, just drag another 'show icon' block onto the screen and attach it to the 'else' part of the 'if, then else' block. Change the icon to anything you like. Now you should find that the simulator changes between the two pictures when you press the buttons.

10

```
on start
    show icon [face]
    set happiness ▾ to 0
```

```
forever
    if   happiness ▾ < ▾ 0
    then    show icon [face]
    else    show icon [face]
```

```
on button A ▾ pressed
    change happiness ▾ by 1
```

```
on button B ▾ pressed
    change happiness ▾ by -1
```

HOW IT WORKS

Congratulations, you have just written your very first computer program. If you like, download the program and copy it onto your micro:bit. If you have a battery pack for the micro:bit you can plug it in and show your friends away from your computer. This is because when you download your program from the Javascript Blocks Editor, it is turned into something that your micro:bit can understand and run. Your program gets stored on a little bit of memory on the micro:bit. Unless you delete it, it'll still be there years from now. Have a go at trying out some of the other blocks in the Javascript Blocks Editor and watch their effect on the simulator.

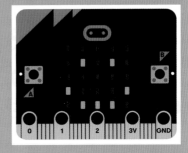

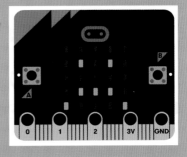

SCUTTLE BOT

In this project, we will use some off-the-shelf motors, known as continuous servos, and a little bit of programming to have a cheeky-looking robot scuttling about in no time.

COMPONENTS

2 continuous rotation servos
1 metal tin (e.g. a mini survival tin)
1 AAA switched battery holder for BBC micro:bit
5 ring electrical crimp terminals (red, 3.2mm)
1 AA 4.5-volt battery holder with switch

1 pair of googly eyes
5 3mm screws with nuts (8mm) (e.g. socket head)
1 roll of double-sided adhesive foam tape
1 BBC micro:bit
1 roll of insulation tape

TOOLS

soldering iron with stand
solder (lead-free)
Allen key (for 3mm socket head)
wire strippers
drilling vice
rotary multi-tool

terminal crimping tool
marker pen
square needle file
superglue
mobile phone or tablet with Bluetooth LE connectivity

SNAPSHOT

Many of our previous robots used small motors to provide motion. They were easy to put together as they only required a battery, a motor and a little bit of wire. The Scuttle Bot is a bit more complex, but if you follow the step-by-step instructions, you'll soon take your skills to the next level.

Motors that are to be controlled from an electronic device like the BBC micro:bit generally require additional pieces of electronic circuitry to be able to switch them on and off, and to change their direction. In this project, we will use a special type of motor that is capable of precise movements – the servo motor.

1 Clamp the metal tin into the drilling vice and position one of the continuous rotation servos about a third from the end (A). Mark the position of the edges of the servo on the tin using a marker pen. Put a cut-off disc into the rotary multi-tool and, using the markings as a guide, cut out a box to allow the servo to fit into the metal tin (B). Take care not to make the cut-out too big as the 'wings' of the servo will be used to secure the motor into place. Repeat this step for the opposite side of the tin. We also need to make a small square-shaped hole in one of the top sides of the survival tin – this should be approximately 1cm (1/3in) wide.

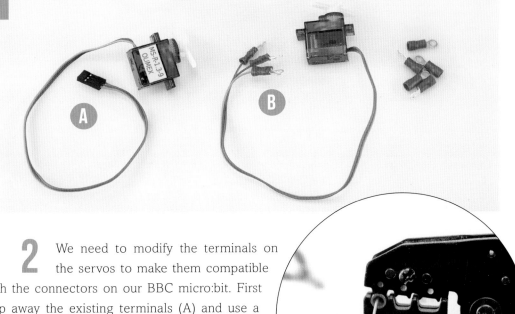

2 We need to modify the terminals on the servos to make them compatible with the connectors on our BBC micro:bit. First snip away the existing terminals (A) and use a pair of wire strippers to expose some bare wire from beneath the insulation. Slip a ring terminal connector over the wire and, using a crimping tool, make a crimped connection (B). Repeat this for all of the wires for each of the servos. If you want, you can use a single crimp for each of the positive (red) and negative (brown) wires, just make sure that the orange wires have their own individual connectors. (If you are going to do this, you must do step 3 first.)

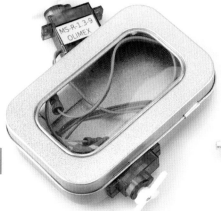

3 Insert a servo into each of the sides of the survival tin and use a blob of adhesive on the servo 'wings' to form a strong bond with the tin.

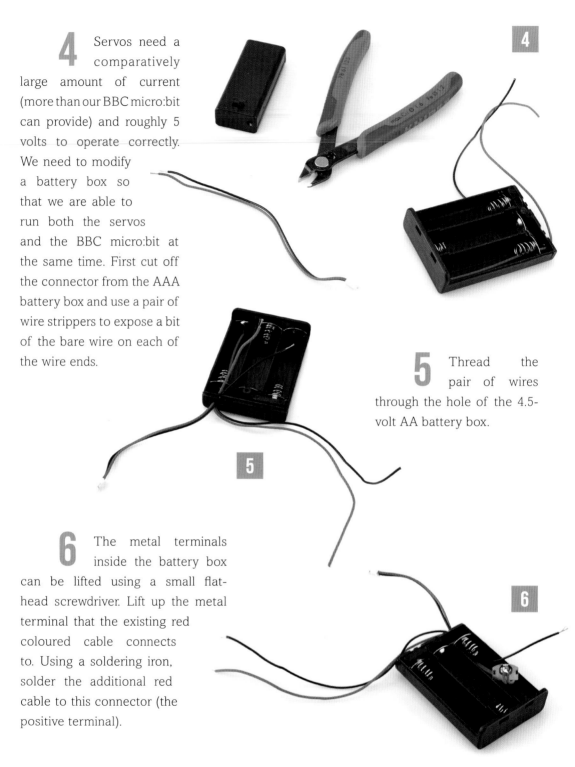

4 Servos need a comparatively large amount of current (more than our BBC micro:bit can provide) and roughly 5 volts to operate correctly. We need to modify a battery box so that we are able to run both the servos and the BBC micro:bit at the same time. First cut off the connector from the AAA battery box and use a pair of wire strippers to expose a bit of the bare wire on each of the wire ends.

5 Thread the pair of wires through the hole of the 4.5-volt AA battery box.

6 The metal terminals inside the battery box can be lifted using a small flat-head screwdriver. Lift up the metal terminal that the existing red coloured cable connects to. Using a soldering iron, solder the additional red cable to this connector (the positive terminal).

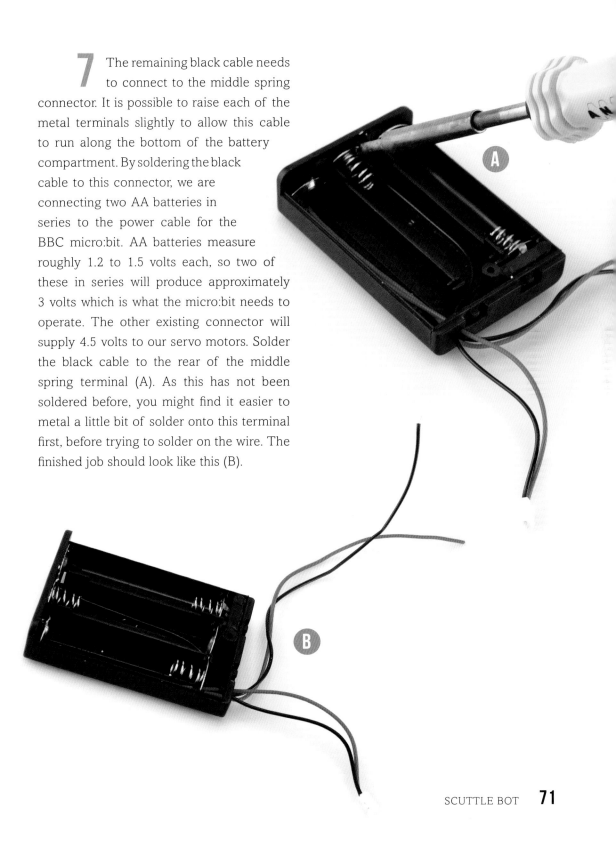

7 The remaining black cable needs to connect to the middle spring connector. It is possible to raise each of the metal terminals slightly to allow this cable to run along the bottom of the battery compartment. By soldering the black cable to this connector, we are connecting two AA batteries in series to the power cable for the BBC micro:bit. AA batteries measure roughly 1.2 to 1.5 volts each, so two of these in series will produce approximately 3 volts which is what the micro:bit needs to operate. The other existing connector will supply 4.5 volts to our servo motors. Solder the black cable to the rear of the middle spring terminal (A). As this has not been soldered before, you might find it easier to metal a little bit of solder onto this terminal first, before trying to solder on the wire. The finished job should look like this (B).

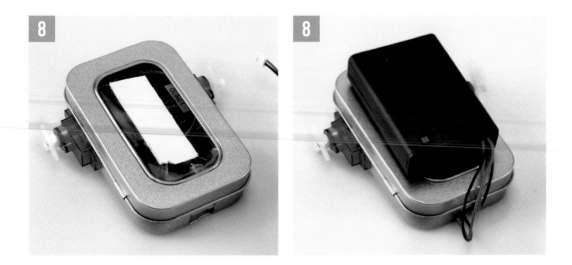

8 Insert AA batteries into the battery box and close the lid. Now apply a bit of double-sided adhesive tape to the top of the metal tin and use this to attach the battery box to the tin. The battery box wires should thread through the hole that we made earlier in the metal tin.

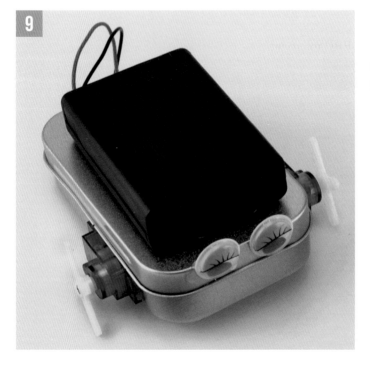

9 Glue the pair of googly eyes onto the unmarked side of the metal tin.

10 Open the lid of the tin and, using the crimping tool, crimp a ring connector onto each of the exposed wires of the battery box.

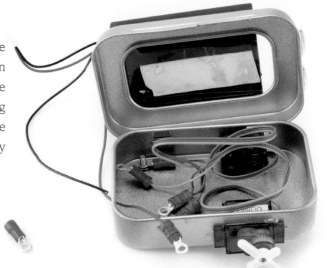

11 We are now ready to wire up our Scuttle Bot. First secure the red (4.5-volt) cables together using a 3mm nut and bolt through the ring connectors (A). Wrap the bare metal of these connectors in a bit of insulation tape to prevent accidental contact with other parts of the robot. Connect the black connector and the two brown cables of the servos to the 'GND' pad of the BBC micro:bit.

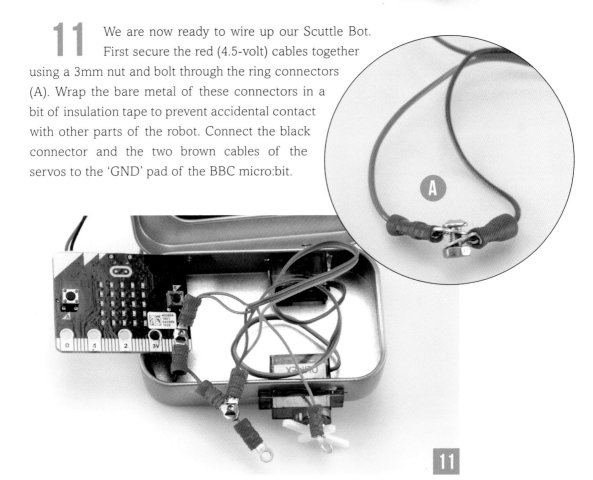

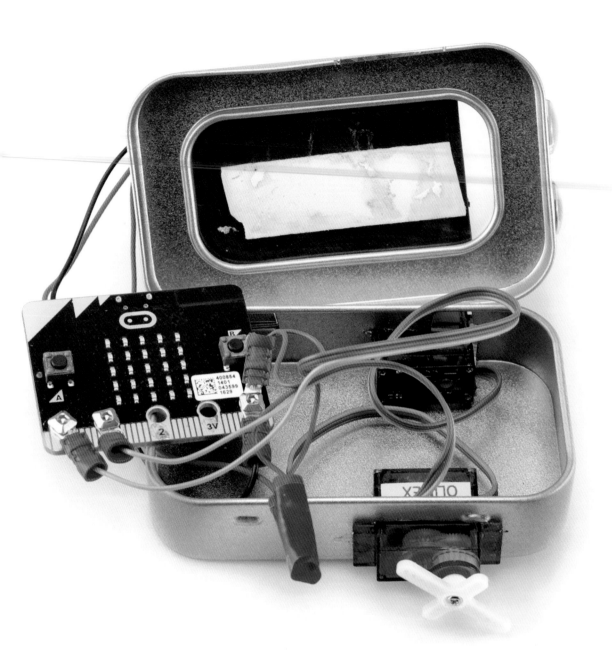

12 Finally connect each of the orange cables from the servos to pad '0' and '1' of the micro:bit respectively.

The hardware for our Scuttle Bot is now complete.

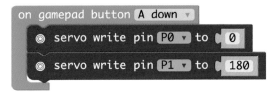
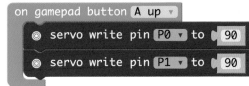

13 We are now going to write a basic program for our Scuttle Bot to allow it to be controlled from a mobile phone. To enable the Bluetooth set of blocks, click on the 'Add Package' tab in the 'Advanced' options and choose 'Devices', accept the warning that we will have to remove the 'radio blocks'. Now click on the 'Devices' tab and drag two 'on gamepad button A down' blocks into the programming area. Change one of the blocks to 'on gamepad button A up'.

Now drag four 'servo write pin P0 to 180' from the 'Pins' tab and put two of these blocks under each of the 'on gamepad button A …' blocks. One of the blocks under each of the 'on gamepad button A …' needs to be changed to P1 – this is because we want to control both of our servo motors.

Finally, we need to adjust the values of the 'servo write pin … to 180' blocks. Continuous rotation servos stop moving when set to 90; values below this to '0' move the servo in one direction and values above this to '180' move the servo in the opposite direction. Set the values to '0' and '180' for each block beneath 'on gamepad button A down' and set the value of the two blocks beneath 'on gamepad button A up' to '90'.

14 Drag a further two 'on gamepad button A down' blocks to the programming area and change these to button 'B'. As before, drag two 'servo write pin P0 to 180' blocks underneath each of these and adjust them as shown.

15

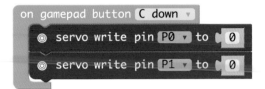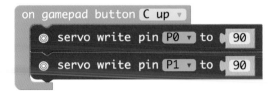

15 Again, drag a further two 'on gamepad button A down' blocks to the programming area and change these to button 'C'. Drag two 'servo write pin P0 to 180' blocks underneath each of these and adjust them as shown.

16

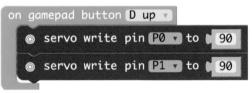

16 Finally, drag a further two 'on gamepad button A down' blocks to the programming area and change these to button 'D'. Drag two 'servo write pin P0 to 180' blocks underneath each of these and adjust them as shown. Phew! No more blocks needed.

Download the program to the BBC micro:bit – make sure that the battery box is switched off and that the micro:bit power cable is unplugged before you do this.

HOW IT WORKS

Unplug your BBC micro:bit from your computer and plug in the battery connections. To control our Scuttle Bot we need two applications on our tablet or mobile phone – the official 'micro:bit' app and the 'micro:bit Blue' app by Martin Woolley. Within the official 'micro:bit' application, follow the instructions to pair your micro:bit with your mobile phone or tablet. Once you have successfully paired your device, you can now load the 'micro:bit Blue' application and choose the 'dpad' application. The left gamepad should now let you control your Scuttle Bot. And away you go!

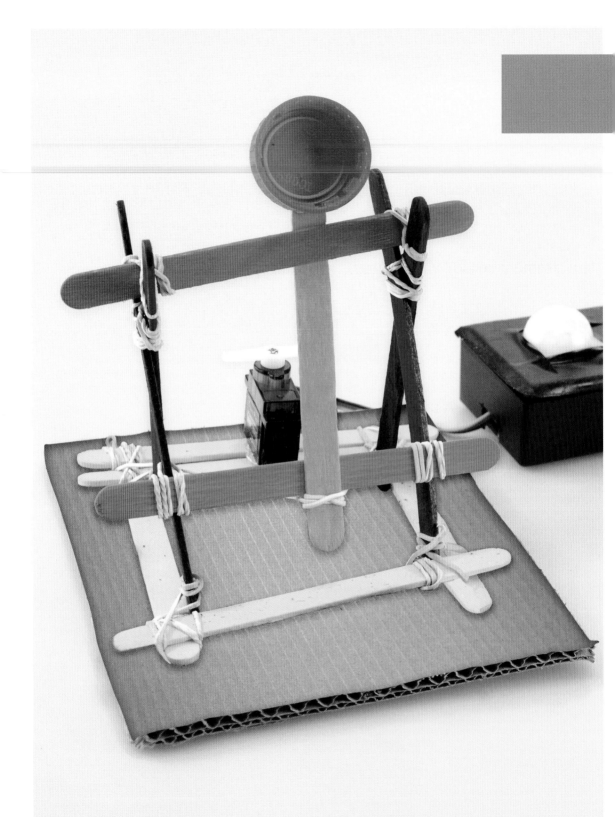

CATAPULT BOT

For this project, we will use an internal feedback loop to control the firing 'pin' of a small catapult. Combined with a simple movement sensor, we can make our catapult fire automatically upon unsuspecting intruders.

COMPONENTS

1 servo
1 project box
1 modified battery box
   (*see* Scuttle Bot)
1 drinks bottle cap
6 ring electrical crimp
   terminals (red, 3.2mm)
5 3mm screws with nuts
   (8mm) (e.g. socket head)

1 BBC micro:bit
1 roll of insulation tape
1 PIR sensor
1 servo tester
1 piece of cardboard
selection of lollipop
   sticks
rubber bands

TOOLS

Allen key (for 3mm socket head)
wire strippers
drilling vice
rotary multi-tool
terminal crimping tool
marker pen
superglue

SNAPSHOT

The Scuttle Bot used continuous rotation servos to let our robot move about under our command. Normal servos are interesting devices; you can make them move to different positions and an internal circuit board and sensor ensures that they move to, and remain at, that position – this is known as an internal feedback loop.

1 Cross two lollipop sticks to form an 'X' shape and wrap an elastic band around the junction to secure the sticks in place. Repeat this step with a second pair of lollipop sticks.

2 Using five lollipop sticks, form a square shape (with two lollipop sticks on one side). Use elastic bands to secure each corner of the square – it helps to first wrap the elastic band around each end of the stick and then around both stick ends.

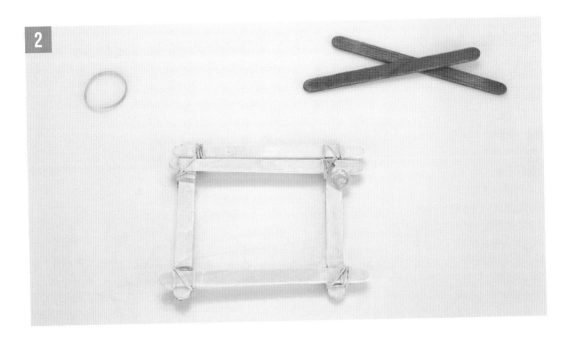

3 Place your pairs of 'X' lollipop sticks on each side of the square and wrap an elastic band around their ends to join them as illustrated.

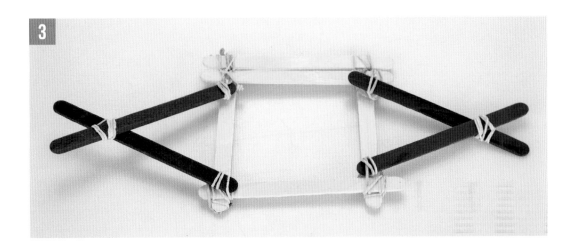

4 Raise the pairs of 'X' lollipop sticks and use two additional sticks to brace them. One stick should be fixed in the internal part of each 'X' and another should brace the bottom part of the frame.

5 Glue the cap of a plastic drinks bottle onto a lollipop stick to form a firing arm. Install the firing arm on the catapult – the bottom part of the lollipop stick needs to be placed in front of the stick bracing the bottom part of the main frame. The top of the stick, near the drinks bottle cap, needs to be positioned behind the lollipop stick that braces the top part of the main frame. Now loop a rubber band around the bottom of the firing arm and stretch this to the back of the catapult. Glue the entire frame onto a piece of cardboard.

5

6 Attach the servo to the servo tester and plug the tester into a 5-volt power source. Select the 'neutral' mode on the tester so that the servo's arm moves to the '90' position. If you do not have a servo tester, you can find this position by rotating the actuator arm to each of its extremes and then moving the arm to the middle point.

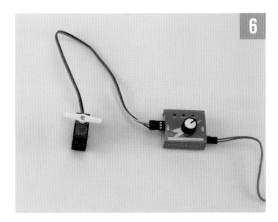

7 Unplug the servo from the servo tester and cut away the connector using snips, then expose a little bit of wire using wire strippers. Crimp a set of ring electrical crimp terminals to each of these bare servo wire ends. At the same time crimp a set of terminals to three additional wires (usually supplied with your PIR sensor). This will let us connect our servo and sensor to our BBC micro:bit.

8 Clamp the project box into the vice and use a rotary multi-tool to cut out a hole for the PIR sensor in the lid. Also create two smaller holes in the front and back of the box, through which we can thread our battery and servo wires.

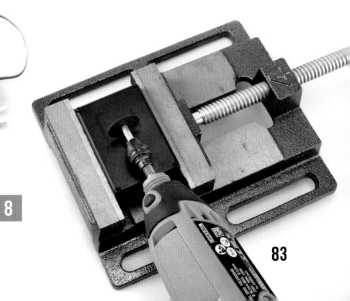

9 We are now going to connect the catapult pieces. Install the PIR sensor into the lid of the box and thread the battery and servo cables through the holes that we just made. Now connect the red cable from the servo to the red lead that is coming in from the 4.5-volt battery box (A); use a nut and bolt to secure these in place and wrap a bit of electrical insulation tape around the connection to prevent it from touching other parts of the robot.

Now tie the black and brown cables from our servo, PIR sensor and battery box to the 'GND' pad on the BBC micro:bit (B). Connect the power cable (C) from the PIR to the 3V pad on the BBC Microbit. Finally, connect the yellow middle cable of the PIR sensor to pad '0' (D) of the micro:bit and the orange cable from the servo to pad '1' (E).

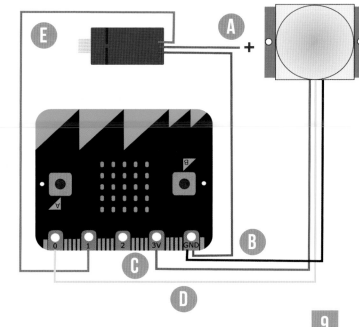

KEY

A   4.5v
B   GND
C   3V
D   PIR Signal
E   Servo Signal

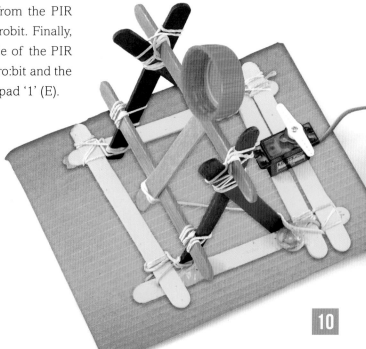

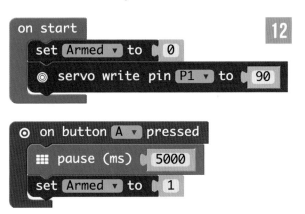

on start   11
  set Armed ▾ to 0
  ◎ servo write pin P1 ▾ to 90

▦ forever

10 Assemble the catapult by fitting the lid on the project box and glue the servo to the catapult. The 'neutral' position of the servo should hold back the firing arm. We are now ready to get on with the programming.

11 Create a new variable in the BBC micro:bit IDE by clicking on the 'Variables' tab and selecting the 'Make a Variable' option. Call the new variable 'Armed'. Now drag a new 'set item to 0' block beneath the 'on start' block and change 'item' to 'Armed'. Drag a 'servo write pin P0 to 180' block (from the 'Pins' tab) beneath this and set the value to 90 , change the pin from 'P0' to 'P1'. This means when we power on our BBC micro:bit it will tell the servo to move to its 'neutral' position.

on start   12
  set Armed ▾ to 0
  ◎ servo write pin P1 ▾ to 90

◉ on button A ▾ pressed
  ▦ pause (ms) 5000
  set Armed ▾ to 1

12 We want a way to arm our catapult so that it doesn't fire as soon as we turn it on. To do this, drag a 'on button A pressed' block to the programming screen (from the Input tab). Drag a 'pause (ms) 100' block beneath this (from the Basic blocks tab) and set the value to 5000. This will mean our program will wait five seconds before executing the next statement. From the 'Variables' tab drag a 'set item to 0' block and place this beneath our wait statement. Change the variable name from 'item' to 'Armed' and change the value to '1'.

```
on start                        13
    set Armed ▼ to     0
    ◎ servo write pin P1 ▼ to    90
```

```
⊙ on button A ▼ pressed
    ▦ pause (ms)    5000
    set Armed ▼ to     1
```

```
⊙ on pin P0 ▼ released
```

13 Drag an 'on pin P0 released' block to the programming screen (the block can be found in the 'Input', 'More' tab). This means when the state of that pin changes, our BBC micro:bit will immediately execute the blocks nested within that statement.

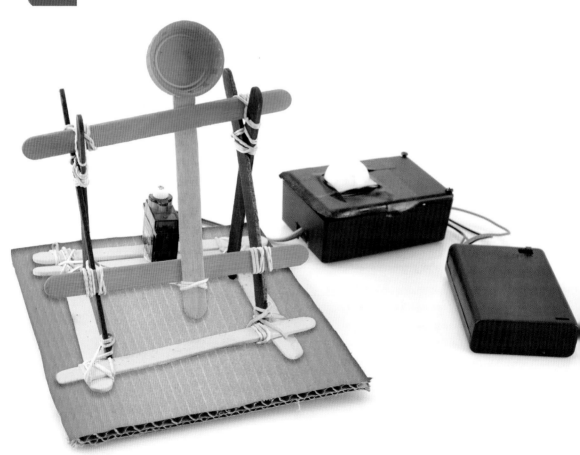

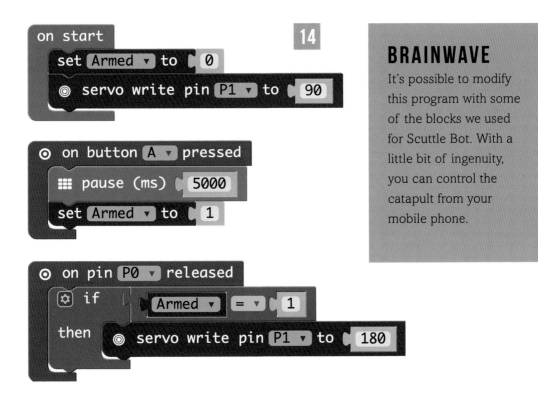

14

BRAINWAVE

It's possible to modify this program with some of the blocks we used for Scuttle Bot. With a little bit of ingenuity, you can control the catapult from your mobile phone.

14 Drag an 'if then' block to the programming screen and connect it to the 'on P0 released' block (the block can be found in the 'Logic' tab). This new block will check to see if our micro:bit has been 'Armed'. To do this, drag from the 'Logic' tab a '0 = 0' block and place this after the 'if' part of our 'if then' block. Now from the 'Variables' tab drag an 'Armed' block to the screen and replace one of the '0's with this block. Change the remaining '0' to '1'. Finally drag a 'servo write pin P0 to 180' block (from the 'Pins' tab) and nest this after the empty 'then' statement.

Our program is now ready to download to our BBC micro:bit.

HOW IT WORKS

Unplug your micro:bit from your computer and plug in the battery connections. Pull back the firing arm and hold it back with the servo arm. When you are ready, press the 'A' button on the micro:bit to arm the catapult. You now have five seconds to take cover! After this time, if the PIR sensor detects any movement it will immediately tell our servo to move its position so that it no longer holds back our firing arm – bombs away!

GARDEN GUARDIAN

In everyday life I'm not much good at keeping my house plants alive because I always forget to water them. However, robots are brilliant at doing mundane, everyday tasks without fail, so perhaps we can harness our robot-building skills to solve this problem.

COMPONENTS

1 metal tin (e.g. a mini survival tin)
2 sprung ballpoint pens
1 large toggle switch
2 googly eyes
1 BBC micro:bit
1 AAA switched battery holder for micro:bit (with batteries installed)

3 ring electrical crimp terminals (red, 3.2mm)
5 3mm x 5mm nuts and bolts
1 soil hygrometer sensor (for Arduino)
1 roll of electrical insulating tape

TOOLS

rotary multi-tool (Dremel)
engineer's pliers
drill vice
crimping tool
wire strippers
superglue or glue gun
marker pen

safety spectacles
needle file (flat)
small flat-head screwdriver
multimeter with continuity mode (useful to have)

SNAPSHOT

For this robot project we will use an inexpensive sensor and a little bit of programming to give our robot the ability to watch over our indoor plants and to alert us when they need watering.

It's common to use a variety of sensors to allow a robot to respond to changes in its local environment. In this project, we will use a sensor that allows our robot to detect the amount of moisture that is present in soil — a hydrometer sensor

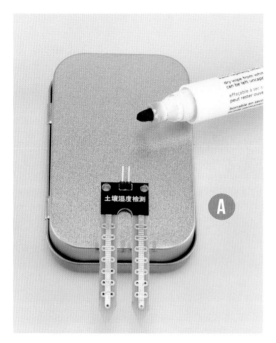

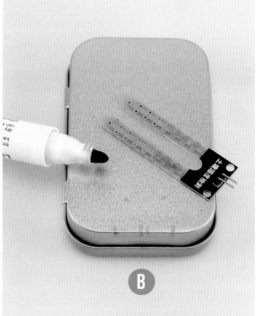

1 Place the resistive soil moisture sensor on the back of the metal tin. Use a marker pen to trace the outline of the hygrometer sensor (A). Extend the outline onto the short side of the metal tin (B).

BRAINWAVE
Make sure that you use a permanent marker pen on the tin or else your hand is likely to rub off the marks that you make.

2 Lay the metal tin flat in a drill vice and ensure that it is held tightly. Insert a cutting disc attachment into your rotary multi-tool. Put on your safety spectacles and plug in the rotary tool. Using the cutting disc, cut a small slot into the metal tin for each of the legs of the hygrometer sensor – be extremely careful as cutting discs can shatter or slip. Don't put too much pressure on the tool while it is cutting (lighter passes are better). It is normal for sparks to be produced while you are cutting into the metal tin. Once you have made the holes, use a flat needle file to deburr the edges.

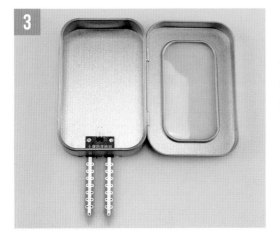

3 Test that the legs of the hygrometer sensor will fit through the holes that you just made. If they do, use the marker pen to mark the sensor's mounting holes onto the tin.

4 Put the metal tin back into the vice and ensure that it is held securely. Swap the rotary cutting disc for a small drill bit in the rotary multi-tool. Drill two pilot holes on each of the markings you just made. Once you have produced these pilot holes, enlarge them with a 3mm drill bit.

5 While the metal tin is still secured in the vice, use the marker pen to trace around the battery holder's power connector – make the tracing on the opposite side to the sensor holes.

6 Re-insert the cutting disc into the rotary multi-tool and cut out a small square hole so that the battery holder's power connector will pass through. Remember, only press lightly on the tool and always wear your safety glasses. Once you have made the hole, deburr it using the needle file.

7 Insert the hygrometer sensor into the metal tin and align the mounting holes. Use the 3mm nuts and bolts to secure the sensor into the metal tin.

You could coat the inside of the tin with electrical insulating tape to help to prevent the metal tin from damaging the electrical components.

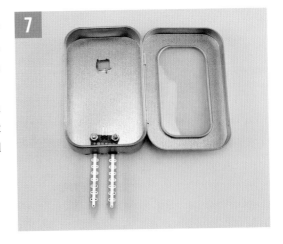

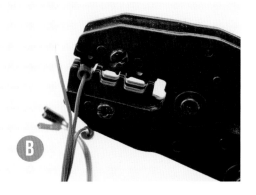

8 The hygrometer sensor should have come with some prototyping wire connectors; however, they are not compatible with the terminals on the BBC micro:bit But don't worry, we roboticists are clever folk. We just need to modify them. First, strip away three wires from the bundle and snip off one end of their connectors using the engineer's pliers (A). Use the wire strippers to expose the bare copper wire. Slip one of the wires through an electrical crimp terminal, the insulation should be inside the crimp's shroud with some wire exposed near the ring side of the crimp. Using the crimping tool, crimp the connector – to do this, put the electrical crimp into the appropriate jaw on the crimping tool and then press down hard to squeeze the crimp (B). The deformed crimp should hold onto the wire tightly. Repeat this step for the other connectors.

9 Dismantle both ballpoint pens to salvage their springs. Springs are extremely useful devices as they can apply a constant force to an object; however, our Garden Guardian doesn't need them for that. Instead, gently pull on each spring to make it a little longer. Put the springs to one side.

10 Take the large toggle switch and glue two googly eyes onto the front; hot glue or superglue is perfect for this. The googly eyes can be glued however you like; personally I like one of the googly eyes to be smaller than the other. It looks wackier.

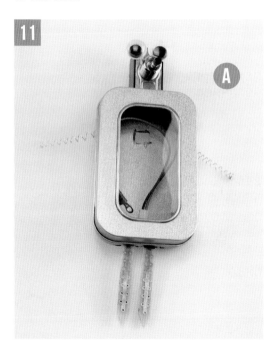

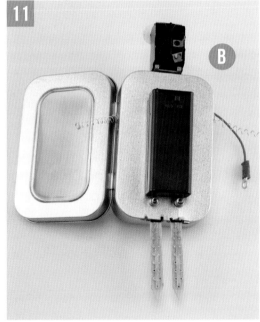

11 Then glue the toggle switch head to the top of the metal tin (on the side opposite the moisture sensor). Attach each spring onto the sides of the metal tin (A). If the parts are giving you trouble, use a little bit of sandpaper to roughen the metal surfaces. Now flip the robot over and glue the battery holder onto its back (B). Our robot is now complete, but we still need to program its brain.

12 Start a new project in the JavaScript Blocks Editor for the BBC micro:bit. To start our program, let's drag a 'show icon' block underneath the existing 'on start' block. The 'show icon' block can be found in the 'Basic' blocks tab. Edit the 'show icon' block by clicking on the default dropdown menu and select whatever image you wish to use. I chose a heart because I want my micro:bit to give a representation of my plant's health.

13 Our hygrometer sensor stops outputting a digital signal whenever the soil moisture drops below a set level. For our micro:bit to be able to detect this we need to drag in an input block. Click on the 'Input', 'More' blocks tab and drag the 'on pin P0 released' block onto the screen. This block will execute the code that is connected to it whenever it detects a low signal on the pin P0.

As our Garden Guardian's purpose is to warn us when our plants need watering, let's drag in some blocks under this to grab our attention. First drag in another 'show icon' block and connect it underneath the 'on pin P0 released' block, we can set the image to anything we like.

We could finish there, but let's also get the Garden Guardian to contact our mobile or tablet. This requires Bluetooth to be available on our mobile or tablet device and for the micro:bit app to be installed. If you have this, in the JavaScript Blocks Editor click

the 'Add Package' tab and choose 'devices'. Accept the warning that we will have to remove the 'radio blocks'. Now click on the new 'Devices' tab, and drag and connect the 'raise alert to display toast' block underneath our existing 'on pin P0 released block'. Toast is not really relevant to our plants, so click on the 'display toast' tab and choose one of the alarms.

14 Currently, our Garden Guardian will display a small heart whenever our plant needs watering – but what if I then water my plants? Currently the small heart will remain. To fix this, drag in a 'on pin P0 pressed' block from 'Inputs' tab. This block will execute code whenever the signal goes high on the pin P0, so let's put another 'show icon' block underneath this. We will change its picture, so that it matches the one that we have under the 'on start' block.

The software is now ready for our Garden Guardian, so download the program that we have just made onto the micro:bit by clicking the 'Download' button and copying the file onto it. If you are on a PC, this should be appearing as a USB mass storage device – if you are on a different platform, follow the guidance provided by the BBC micro:bit website on how to do this.

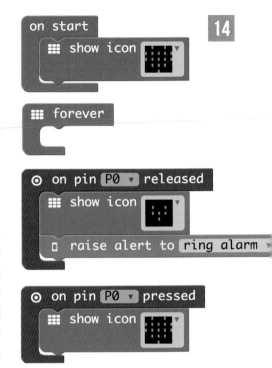

HOW IT WORKS

Switch on the battery pack and put the Garden Guardian into a plant pot. Open the front of the tin and use a small flat-head screwdriver to adjust the little dial (potentiometer) on the hygrometer breakout shield. This dial adjusts the threshold at which the shield will emit a digital signal; keep adjusting this until the screen changes from the full heart to the small heart picture. When this happens, irrigate the soil with a little bit of water – you should find that the alarm stops and the LED display switches back to the full heart picture.

This happens because the electrical resistance of the probes changes depending on the amount of moisture in the soil. The small breakout board that is connected to these probes is looking at the resistance and comparing it against a set value (selected using the little dial). Depending upon the output of this comparison, the breakout board may or may not emit a signal on the 'digital out' pin.

15 Our Garden Guardian is nearly ready to protect our plants. So, let's finish its assembly. Attach the breakout shield (A) that comes with the hygrometer sensor to the sensor probes by using the spare prototyping cables (it's the side with only 2-pin headers). On the other side, attach a crimped cable to each of the terminals labelled: VCC, GND (B) and DO (C). The ring electrical crimp should be bolted to the micro:bit terminals. VCC goes to the terminal labelled '3V' (D), GND goes to the terminal labelled 'GND', and DO goes to the terminal labelled '0' (*see* diagram below).

Put the micro:bit into the metal tin and attach the battery power connector wire to the battery input connector on the micro:bit (it's the connector on the top right). Our robot is now complete and ready to protect our indoor plants.

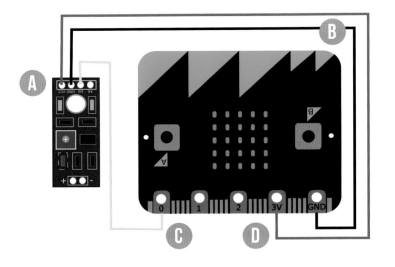

KEY

A Breakout shield
B GND wire
C DO
D 3V/VCC wire

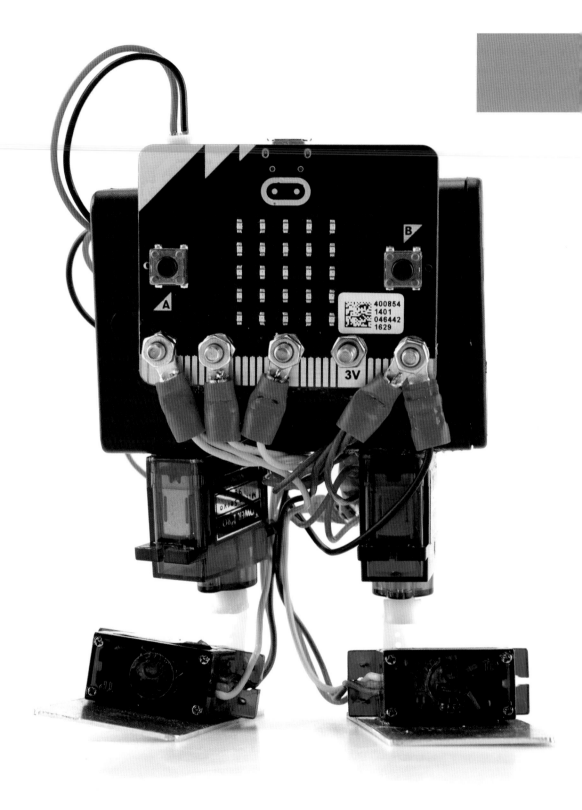

WALKING ROBOT

For our final intermediate project, we will construct a basic walking robot. It demands a bit of ingenuity, but once up and running we will be giving Asimo (an expensive robot built by Honda) some healthy competition in no time.

COMPONENTS

4 servos
1 modified battery box (*see* Scuttle Bot) with AA 1.5-volt batteries installed
6 ring electrical crimp terminals (red, 3.2mm)
5 3mm screws (12mm length) with nuts (socket head)
1 BBC micro:bit
1 roll of insulation tape
1 servo tester
2 dog tag blanks

TOOLS

Allen key (for 3mm socket head)
wire strippers
helping hands
Sugru mouldable glue or a lighter
terminal crimping tool
superglue

SNAPSHOT

For humans, walking is an instinctive action to which we don't give a second thought. However, for robots, walking is very difficult. In recent years, some very expensive robots have been built that can walk, run and even climb stairs – but despite all of their advances, they still have a habit of toppling over!

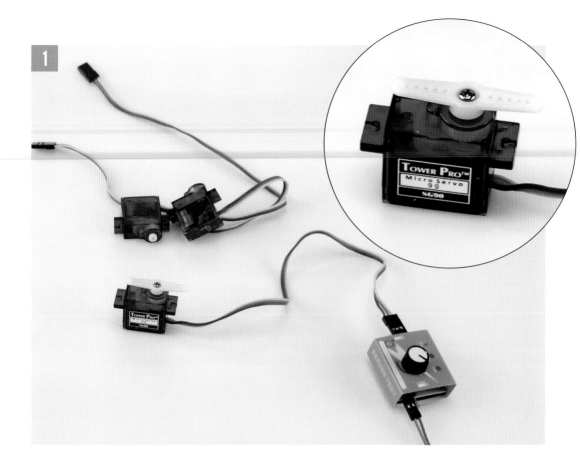

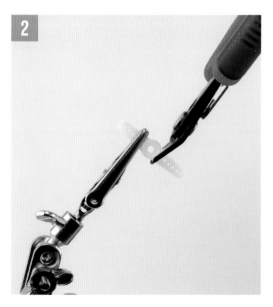

1 Plug each of the servos into the servo tester and set them to their 'neutral' position. If you don't have a servo tester, it is possible to find this position by moving the servo arm to each of its extreme positions and then moving it back to the middle point.

2 Carefully remove the servo arms from the servos and grip one in a pair of helping hands. Cut off the end of the arm so that the part that connects to the servo only remains.

3 Remove this piece (A) from the helping hands and replace it with a second arm (B). We now want to attach the piece that we just cut off to this servo arm to form a joint for the legs. There are two ways to do this:

1) Mix some Sugru (a readily available mouldable putty) and sculpt a joint. You can temporarily secure the pieces with a small amount of superglue until the Sugru hardens.

2) Carefully use a gas lighter to melt the two pieces together. This method is inexpensive and creates a very strong bond. You should ask an adult to do it for you. Create two separate leg joints using one of the above methods.

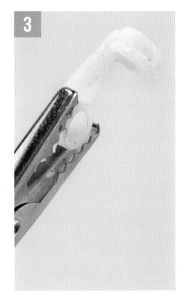

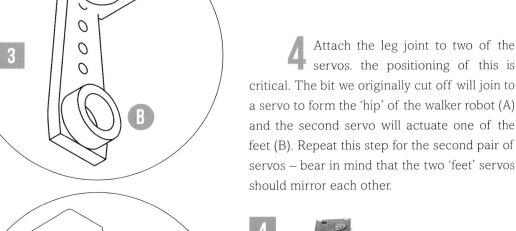

4 Attach the leg joint to two of the servos. the positioning of this is critical. The bit we originally cut off will join to a servo to form the 'hip' of the walker robot (A) and the second servo will actuate one of the feet (B). Repeat this step for the second pair of servos – bear in mind that the two 'feet' servos should mirror each other.

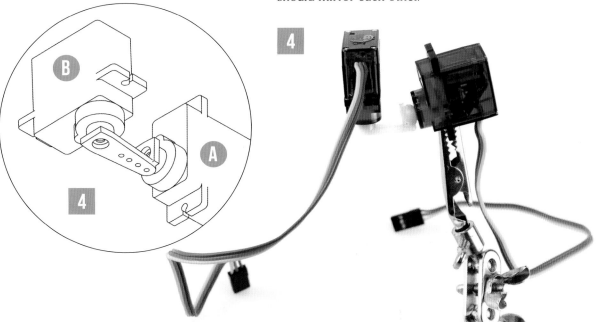

5

5 Cut off the connectors to the servos and crimp on ring electrical crimp terminals. To save on connectors, all of the red leads can connect to one terminal. Similarly, all of the brown leads can connect to a second terminal. The yellow leads of the hip servos can also be joined together, but the remaining two yellow wires of the feet servos must have their own individual crimped connections.

6 Glue the feet servos to the dog tag blanks using some superglue. You should place the servos near one end of the dog tag blanks – the servo arms face towards the back part of the foot. The feet should be able to keep themselves upright, if not adjust the balance and weight of each foot. Put the servo motors to one side.

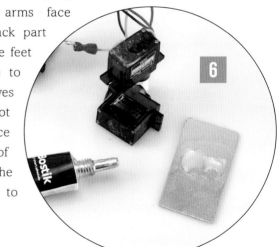

6

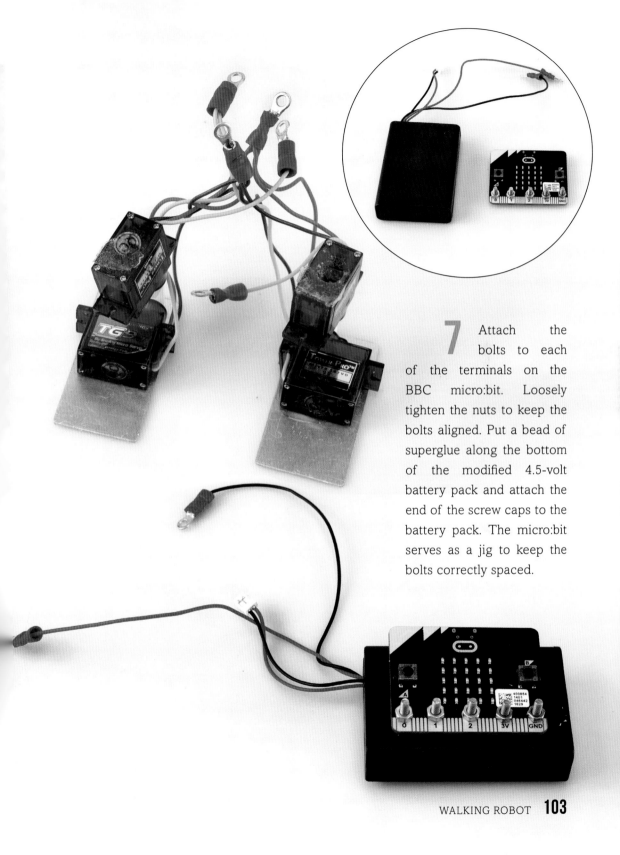

7 Attach the bolts to each of the terminals on the BBC micro:bit. Loosely tighten the nuts to keep the bolts aligned. Put a bead of superglue along the bottom of the modified 4.5-volt battery pack and attach the end of the screw caps to the battery pack. The micro:bit serves as a jig to keep the bolts correctly spaced.

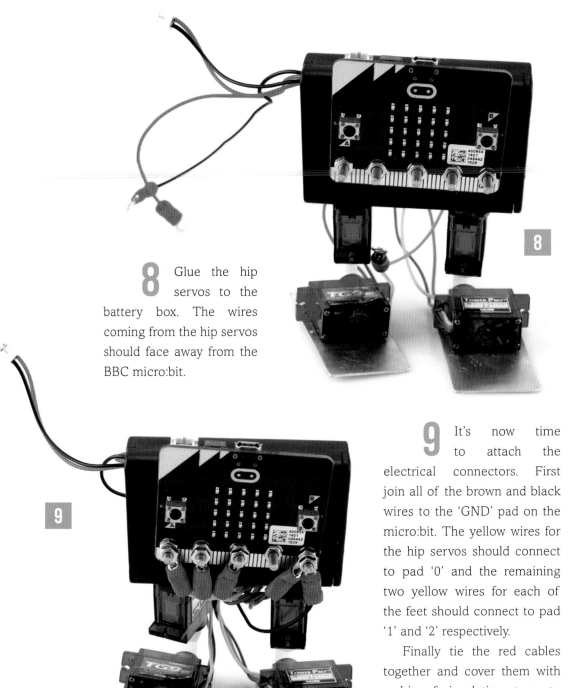

8 Glue the hip servos to the battery box. The wires coming from the hip servos should face away from the BBC micro:bit.

9 It's now time to attach the electrical connectors. First join all of the brown and black wires to the 'GND' pad on the micro:bit. The yellow wires for the hip servos should connect to pad '0' and the remaining two yellow wires for each of the feet should connect to pad '1' and '2' respectively.

Finally tie the red cables together and cover them with a bit of insulation tape to prevent them touching other components.

Our physical robot is now complete – now to program it.

10

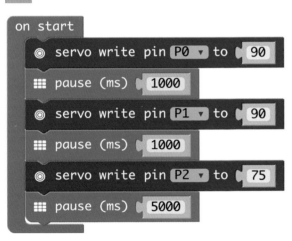

```
on start
    ◎ servo write pin P0 ▾ to  90
    ::: pause (ms)  1000
    ◎ servo write pin P1 ▾ to  90
    ::: pause (ms)  1000
```

11

```
on start
    ◎ servo write pin P0 ▾ to  90
    ::: pause (ms)  1000
    ◎ servo write pin P1 ▾ to  90
    ::: pause (ms)  1000
    ◎ servo write pin P2 ▾ to  75
    ::: pause (ms)  5000
```

10 When our program starts, we want all of our servos to move to their 'neutral' position. We do this by using the value '90' with a 'servo write pin' block. Because it takes time for servos to move from one position to another, we need to use 'pause' blocks to let our robot physically move. To begin our program, use 'servo write pin P0 to 180' blocks to set our servos to their home position. Adjust the values from 180 to 90 and make sure that each 'servo write' block is either setting 'P0' or 'P1'. Add 'pause' blocks (from the Basics tab) and set them to 1000ms as shown.

11 We now need to do the same for our other servo motor. Again, use a 'servo write pin P0 to 180' block and adjust it to refer to 'P2'. You will notice that instead of '90' I have chosen '75' as the value to write to the servo. In your case, you might start with the value '90', but if on start-up you find that your walker immediately tips over, be prepared to tinker with this value.

Follow this block with a 'pause' block and set its value to 5000 – this way our robot should wait for five seconds before attempting to walk; giving us a chance to make sure the initial setup of our walker is right.

12

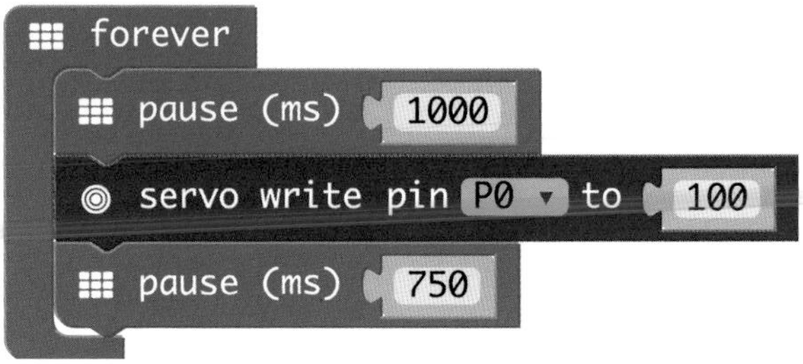

```
forever
    pause (ms) 1000
    servo write pin P0 ▾ to 100
    pause (ms) 750
```

12 As with the start of our program, each motor movement must be followed by a 'pause' statement to provide the servo motors adequate time to complete their movement. Study the pictures to see how the 'pauses' are built in. I have chosen '750' as a good middle ground to provide the motors time to move, but not too long to get boring!

As the picture shows, insert a 'servo write pin P0 to 180' block and set its value to '100'. This block should cause the hip to move.

13

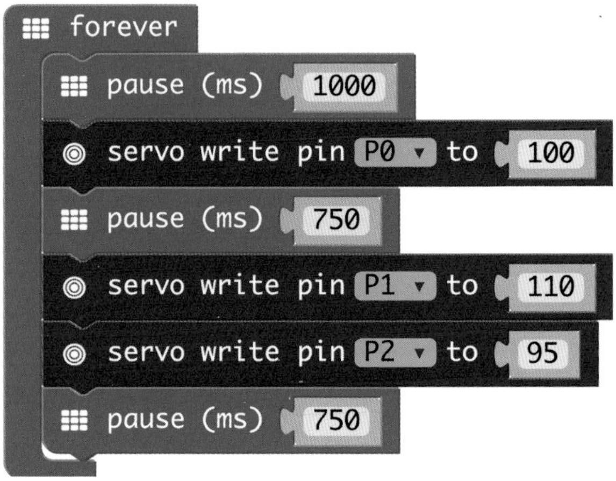

```
forever
    pause (ms) 1000
    servo write pin P0 ▾ to 100
    pause (ms) 750
    servo write pin P1 ▾ to 110
    servo write pin P2 ▾ to 95
    pause (ms) 750
```

13 Following this block, we now want to move both of our feet servos at the same time – this way our robot will unbalance itself and start to move forward by shifting its feet to take a step. Again use 'servo write pin P0 to 180' blocks for this and set them to 'P1' and 'P2'. I have chosen the values '110' and '95', followed by a 'pause' statement.

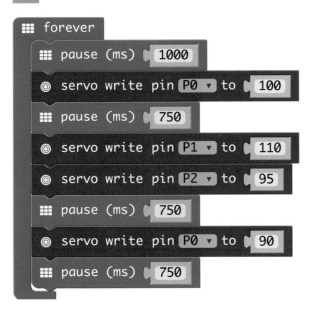

14 We now want our robot to put its feet back down. To do this we need to set the hip position back to '90'.

15 Now we'll take another step, but this time with the other foot. So, use a 'servo write pin P0 to 180' block to move the hip; I set the value '80' this time to move the hip in the other direction.

```
forever
    servo write pin P0 ▾ to  100
    pause (ms)  750
    servo write pin P1 ▾ to  110
    servo write pin P2 ▾ to  95
    pause (ms)  750
    servo write pin P0 ▾ to  90
    pause (ms)  750
    servo write pin P0 ▾ to  80
    pause (ms)  750
    servo write pin P2 ▾ to  55
    servo write pin P1 ▾ to  70
    pause (ms)  100
    servo write pin P0 ▾ to  90
```

16

16 Again, we will move both feet at the same time, so use 'servo write' blocks and set their values to '55' and '70' – this will move the feet in the opposite direction. Finally return the hip to its neutral position by setting its position to '90'.

It's taken some time but now our program is complete.

BRAINWAVE

If the walker is really struggling to maintain its balance, even when it's turned off and the servos are in their neutral position, a small weight on each foot can help to counterbalance it.

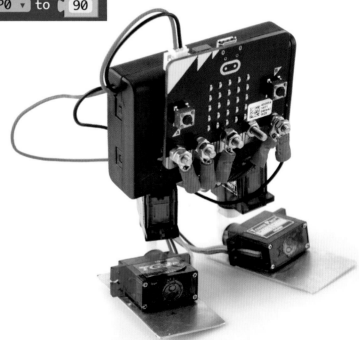

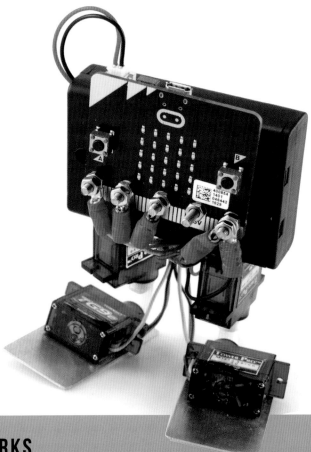

HOW IT WORKS

While the battery pack is switched off and the cable unplugged, download the program to the micro:bit. Place the walker on a flat surface and switch on the power to the motors. When you are ready, plug the battery pack into the micro:bit's power connector. The servo motors should move to their home position (with a bit of luck this will be roughly the same as their current position). If the robot immediately topples over, adjust the values in the start-up part of the program and repeat. If it stays upright, within the next five seconds our walker should attempt to take its first step into the big wide world. Just like a child, this will probably be quite unsteady, so be ready to catch it! You will need to gradually adjust the values to the servos at each stage to help find the right position to maintain the walker's centre of gravity. Don't be discouraged if it keeps falling over, even the best walkers have to learn step by step. Once tuned, your robot should happily (albeit somewhat haphazardly) walk around all day long.

03

SMART MAKES

ROBO WARRIOR

CNC WRITER

MARS ROVER

These advanced projects introduce you to the world of programming. Follow the simple instructions to create some truly impressive home-made robots.

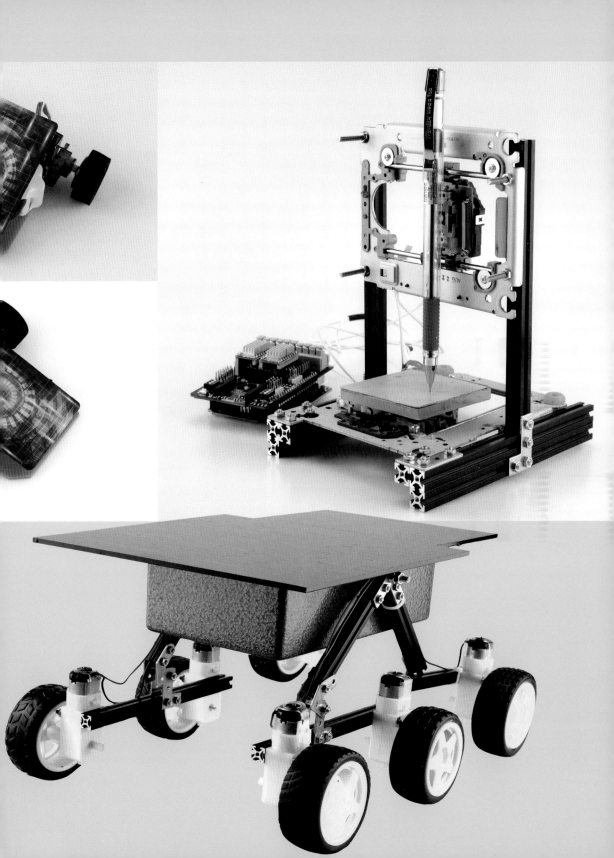

ROBO WARRIOR

A few years ago a new sport was born – robot combat. The idea was that contestants would build large fighting robots and let them battle it out in an arena. This may seem technically very difficult to do, but this project shows you how to build a robo warrior from simple components.

COMPONENTS

1 9g servo
2 continuous rotation servos
2 wheels for servos
1 6-volt 210mAh ultra-miniature
   1/2AAA NiMH battery with Futaba
   connector
1 R/C switch battery receiver on-off
   Futaba JR connector leads

1 expired gift or credit card
1 laser-cut panel or piece of thick
   cardboard (*see* diagram page 114)
1 Hobby King HK-T6A V2 transmitter
1 Hobby King HK-T6A V2 receiver
1 Hobby King 2.4Ghz 6Ch Tx USB cable
   (to download firmware)
8 AA batteries

TOOLS

drill vice
rotary multi-tool with cut-off disc
hot glue gun
marker pen
flat-head screwdriver

SNAPSHOT

For this project, we will put together a simple, cheap and lightweight fighting robot that can be modified and upgraded as your skills develop.

In recent times electronic parts have been getting cheaper and simpler to work with, and so it has become much easier to construct these kinds of fighting machines. While the battle bots that you see on TV shows like Robot Wars are extremely advanced, you can still participate in the sport by joining one of the smaller leagues.

In the UK one of the classes is known as 'Antweight' for which the weight limit is 150 grams (5¼ oz). There are still quite a few rules to observe (they can be found on the Internet), but this weight class is much more approachable for the beginner, and some of the robots that people have built are just as devastating.

BRAINWAVE

It is important to note that, while fun, combat robots can potentially be dangerous – so please use caution if you decide to construct your own.

ROBOT BASE TEMPLATE

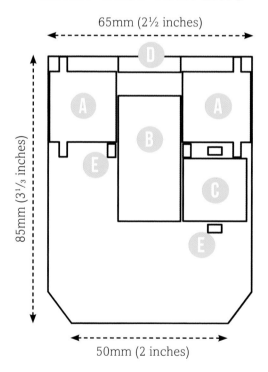

65mm (2½ inches)

85mm (3⅓ inches)

50mm (2 inches)

1 We first need to produce the base for our robot. If you have access to a laser cutter, cut the following DXF file (https://github.com/danielknox/Robot_Warrior) out of 3mm acrylic or wood. The file is calibrated in millimetres – the red lines represent cuts and the blue lines engravings. If you don't have a laser cutter, not to worry. Just transcribe the drawing (left) onto a piece of stiff card and cut it out.

KEY

A	Continuous rotation servo
B	HK-T6A receiver
C	NiMh battery
D	R/C switch
E	Holes for zip ties

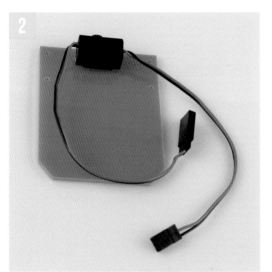

One of the important rules for antweight robots is that they must have an accessible removable link (a piece of wire that can be detached) or an On/Off button that can disable the robot. We will use a simple R/C switch to cut all power to the electronics. Attach the R/C switch to the rear of the plastic plate using a hot glue gun (annotated D on the diagram opposite) – the switch should face outwards.

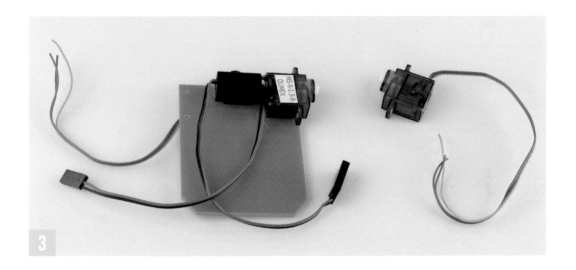

We are now going to attach the servos that will provide our robot with motion. To do this, attach both continuous rotation servos to the back of the plastic plate (annotated A on the diagram opposite) using hot glue. The servos should sit on either side of the R/C switch. Continuous rotation servos are simple to work with, but many more experienced roboteers use small metal gear motors with higher rpms (they make the robot move faster).

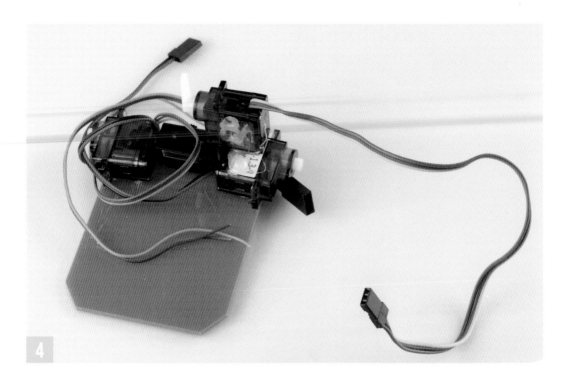

4

Now it's time to build our robot's deadly weapon. Attach a servo horn to the 9g servo and check its rotation. We want to place the servo facing upwards with the horn able to move between a 10 o'clock position and a 2 o'clock position. Once you have the horn in the right spot, hot glue the servo in place near the front of the right-hand side servo.

5

Clamp the gift or credit card in a drill vice and mark the location for a small slot near the top centre part of it. Using a rotary multi-tool with a cut-off blade, create a small slit in the card that is just wide enough for the servo horn to fit through.

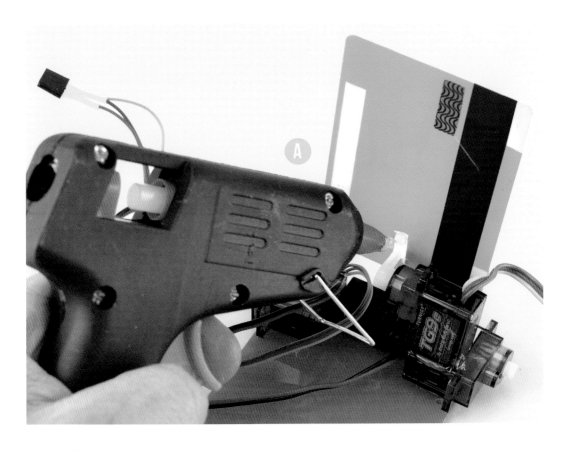

6 Slip the plastic card over the servo horn of the combat robot. Rotate the servo horn with the card in place until the horn points away from the robot; the card should point up towards the sky.

Once you have aligned the card, hot glue it in place; a little dab of glue on the underside of the card helps to tack it in place (A). You can then secure it firmly with a nice dollop of glue on the top (B).

When you activate the servo, the credit card part of the robot will flip up. This will allow you to topple opponents onto their backs when you have manoeuvred the Robo Warrior close enough to them.

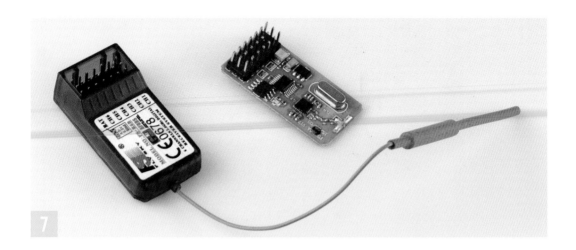

7 The RC receiver is a bit big for our purposes because of its plastic casing. Carefully remove the casing using a flat-head screwdriver to expose the small electronic board.

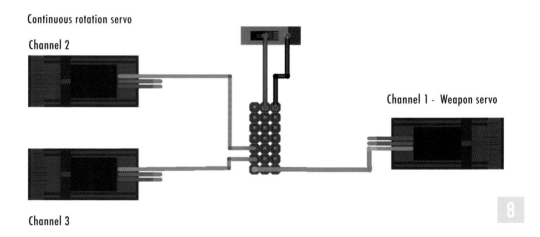

Continuous rotation servo

Channel 2

Channel 1 - Weapon servo

Channel 3

8 Referring to the diagram on page 114, position the NiMH battery (annotated C) and receiver board (annotated B) on the plastic base and secure these in place with zip ties (annotated E). If you are using a bit of plastic, you can hot glue the receiver and battery pack in place. Now plug the continuous rotation servos into channels 2 and 3 of the receiver board, the 9g servo into channel 1 and finally the output of RC/switch cable into row 7 – this was labelled BAT when the casing was in place. Tidy the wires up into bundles using zip ties.

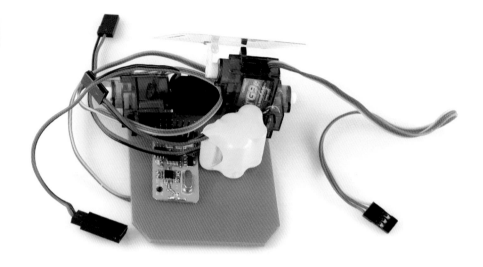

9 Attach the wheels to your robot; it's up to you what diameter wheels you use – I prefer small diameter ones. Once the wheels are secured, our robot is now physically ready to fight.

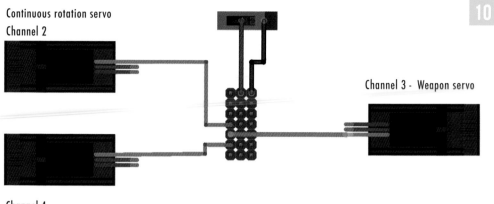

Continuous rotation servo
Channel 2

Channel 3 - Weapon servo

Channel 4

10
By default, the steering on our robot is controlled by the left and the right sticks of the remote control. Some people find this type of steering a bit unnatural, especially when used in conjunction with the flipper, as it can be difficult to steer and operate the weapon at the same time. However, it's possible to reprogram the transmitter, using a programming cable, to adjust the controls to a different set-up; this is known as mixing. First we need to physically change the wiring on our receiver – the continuous rotation servos should move from channels 2 and 3 to channels 2 and 4. The servo for the flipper should now move to channel 3.

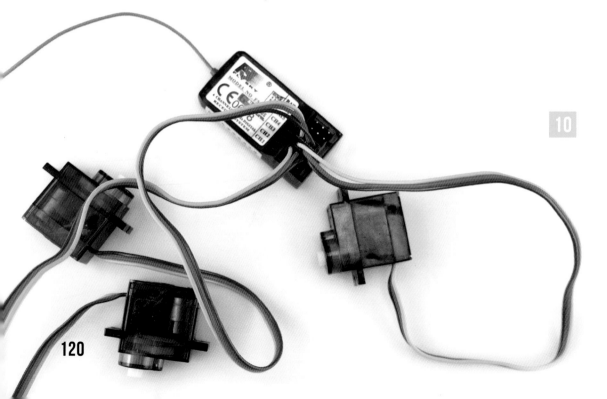

11 We need to install the drivers for the Hobby King 2.4Ghz 6Ch Tx USB cable before we plug it into a computer. Download and install the Silicon Labs CP210x USB to UART Bridge drivers (http://www.silabs.com/products/development-tools/software/usb-to-uart-bridge-vcp-drivers); you may need to restart your computer once these drivers have installed. You can now plug the adapter cable into your transmitter, turn on the transmitter and plug it into your PC. To check that the device has been detected correctly on Windows computers, open 'Device Manager' and look under the 'Ports' dropdown. If you see the Silicon Labs CP210x USB to UART device listed, the driver has installed correctly. Make a note of the Com port number. On Mac (OSX), you still need to install the driver, but you don't need to check 'Device Manager'.

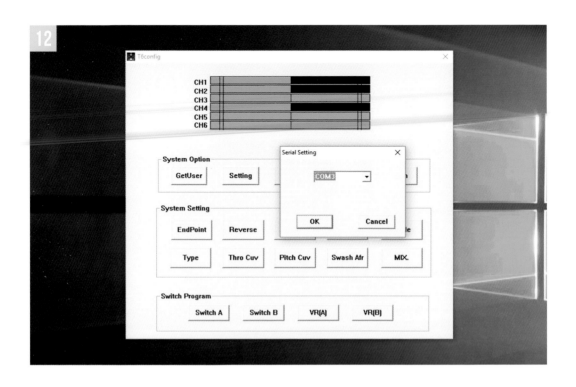

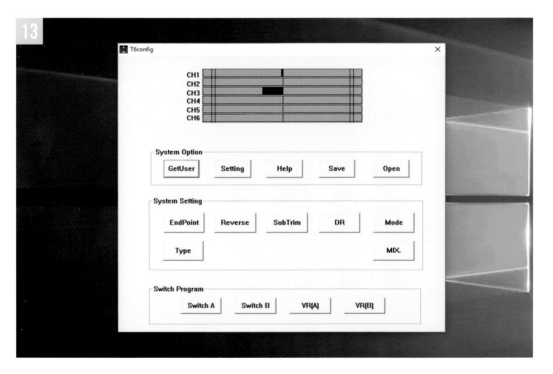

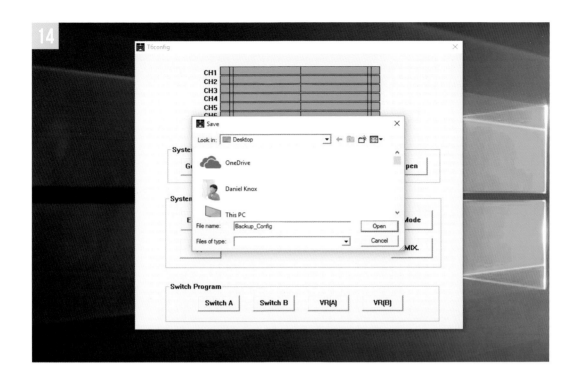

12 The next steps only apply to Windows computers. If you are on a Mac, you can use the program 'TurborixConfig' (http://www.zenoshrdlu.com/turborix/) to configure your controller. On Windows install and open T6config (https://hobbyking.com/media/file/824762106X399553X54.zip) and click the 'Setting' button. The 'Serial Setting' dialog opens and from the dropdown box select the Com port that matches the number you just noted down. Click OK.

13 Test that the controller has successfully connected to your computer by turning some of the sticks on the transmitter – you should see the green bars in the T6config software also move.

14 It's a good idea to make a backup of the factory settings before we modify them or write our own. Click 'Save' and enter a filename; for example, type 'factory_backup'. Click 'Open' to save the file.

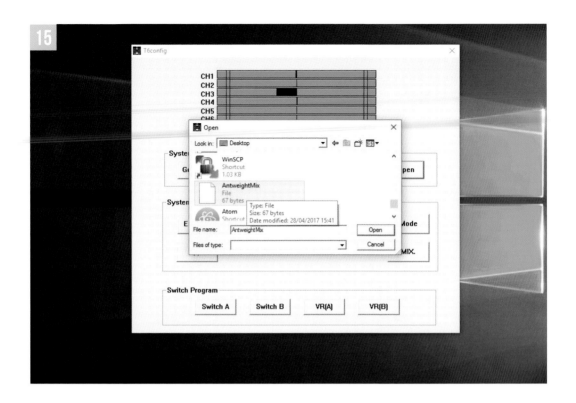

15 We are ready to write our new mixing settings to the transmitter. First download the mix file from (https://github.com/danielknox/Robot_Warrior/tree/master/Software) to your computer. Now in T6Config click 'Open' and then navigate to the directory containing the file that you just downloaded. Select the file and then click 'Open'. You should now be able to move the sticks on the transmitter. The right-hand stick should control the direction of channels 2 and 4, and the left-hand stick should control channel 3.

You are now ready for combat!

BRAINWAVE

Remember that even a relatively low-powered and simple robot can do extremely well in a fight – the secret is to become really well practised at knowing how to control and manoeuvre your robot (A, B). Like David and Goliath, a small but skilful fighter can take down even the biggest and scariest opponents.

HOW IT WORKS

We are now ready to operate our tiny antweight robot. First charge the NiMH battery using an appropriate battery charger. Once charged, plug the battery into the R/C switch input. When you are ready to do battle, switch the robot on and turn on your transmitter. Although our robot moves about fairly slowly, it can push something quite powerfully (C). The onboard weapon will help you flip over opponents (D), but it takes some skill to self-right yourself with it if you turn over (E). At the end of your fight or practice session, don't forget to switch off the robot using its R/C switch.

CNC WRITER

CNC-enabled machines allow us to perform repetitive jobs at high speed and with great precision. CNC (computer numerical control) can seem a bit like magic at first, but building a computer-controlled machine, and operating it, isn't as hard as you might imagine.

COMPONENTS

M3 square head set bolts x 6mm (pack of 250)

M3 square head set bolts x 25mm (pack of 4)

M3 stainless steel plain nuts (pack of 250)

2 MakerBeam – 40mm (1½in) long black anodized beam

4 MakerBeam – 150mm (6in) long black anodized beam

Arduino Uno microcontroller board

DFRobot dual bipolar stepper motor controller for Arduino

9-volt power supply

2 second-hand CDROM/DVD drives

4 90-degree-angle L bracket

22 AWG wire (stranded)

small piece of acrylic

roller pen

TOOLS

soldering iron

lead-free solder

snips

3mm nut driver (5.5)

wire strippers

marker pen

superglue

multimeter

flat-headed screwdriver

SNAPSHOT

All it takes are two reclaimed CDROM/DVD drives, a controller board and some simple extrusion, and we'll have a cool-looking CNC machine put together in no time.

Most CNC machines use steppers/servos to control their movement allied to some form of gantry/rail system – accurately assembling this can be quite hard. Luckily for us, mass-production means there's a ready-made system conveniently to hand: old CDROM/DVD drives.

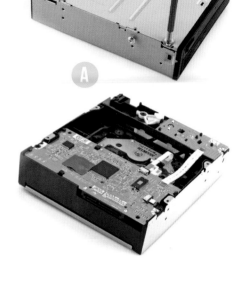

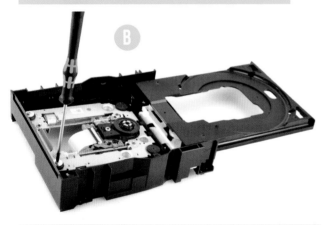

1 We first need to salvage the motion drive from our two CDROM/DVD drives. This can be a little fiddly and is unique for each brand of drive. Essentially, undo all of the screws in the drive. Remove the front face of the drive, which is usually held in place by plastic tabs (A). You should now be able to open up the drive and see inside. Remove all the screws holding the daughter boards and detach the ribbon cables, held in place by small plastic clips (B). With most of the boards out of the way, you should now be able to lift out the metal motion carriage (C).

2 Stepper motors are an interesting breed of motor – unlike servos they do not have a feedback system to find their current position, but they have more control than standard DC motors. They operate by rapidly pulsing the windings of the motor, in a controlled manner. A stepper motor controller will do this complicated step for us, but first we need to identify the pairs of coil windings in the motor.

Most drives are bipolar motors, so two of the contacts will connect to one winding and the other pair will connect to the second winding. To identify the pairs, we will use a multimeter on the ohm (Ω) measurement setting. Using the multimeter probes, test two of the solder contacts on the stepper motor. You should get an ohm readout from each of the contacts. What we are looking for is a value less than infinite, but more than 0 – this is because the contacts that are not connected to each other will read infinite resistance, whereas the contacts that are connected to each other will get a small resistance reading.

Once you have identified the two pairs, label them (this becomes very important later!).

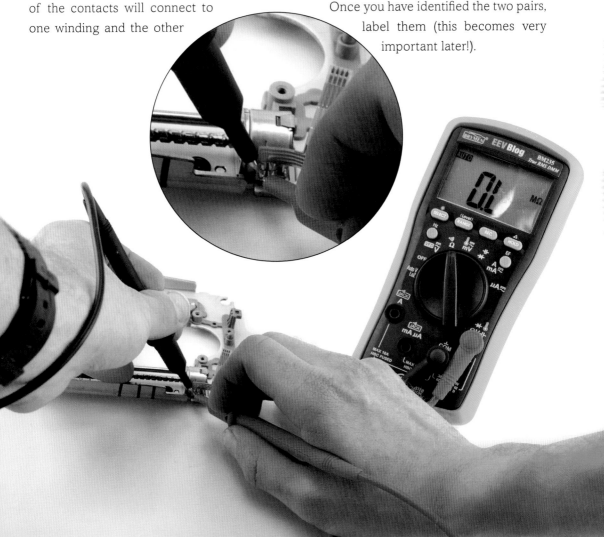

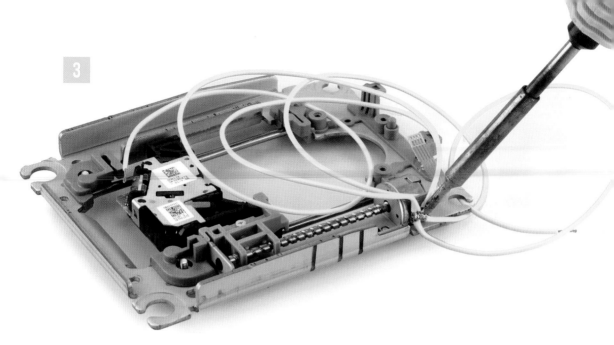

3

Now that we have identified the pairs of windings, solder wires onto each of the contacts. You can solder onto the ribbon cable if you like, or directly onto the motor solder joints. It is important that the solder joints and wires do not bridge – this can be a little fiddly and it helps to have pre-tinned wires. Once the wires are on, it's a good idea to label each cable with the motor and pair number.

4

We will be assembling the CNC frame using aluminium extrusion. The particular type used here is known as Makerbeam; which gives us a tiny 10mm x 10mm profile; we will also use special square bolts and a nut driver to tighten the components. Each CDROM/DVD drive is a little different, so you must first look carefully at each drive to work out how best and where to mount it to the extrusion.

First slip two 25mm (1in) square bolts into a groove of a 150mm (6in) piece of extrusion. Secure these bolts in place with a M3 nut. Repeat this step.

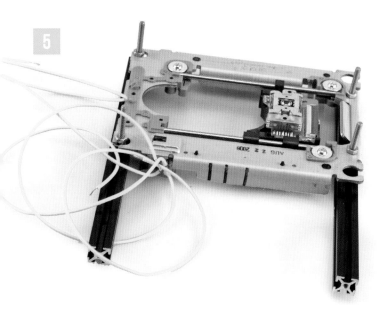

5 Now mount one of the drives to this frame. This will form the X axis. You will see in the picture that I chose to use two holes on the right-hand side of the drive. For the left-hand side, grip the two original mounting points between a pair of nuts. Once you have found the best way to mount the drive to the top of the frame, tighten everything into place.

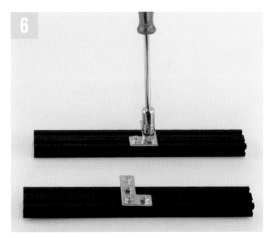

6 We are now going to construct the bottom frame of the CNC machine. First use one of the L brackets to attach two pieces of 150mm (6in) extrusion together. Also create a mirrored copy for the other side of the machine.

7 Use another L bracket to attach a 40mm (1½in) length of extrusion to the top piece of the extrusion. This will form the front part of the CNC machine. As before, create a mirrored copy with the other piece of frame.

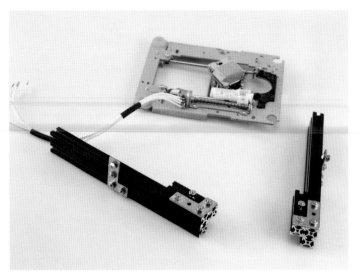

8 To form the Y axis, attach the second CDROM/DVD drive to this bottom frame. Just like the last frame, I'm using a nut to grip the drive's mounting points (A). The shape of this particular drive means that it is supported securely by the extrusion, so you don't need to worry about adding additional support.

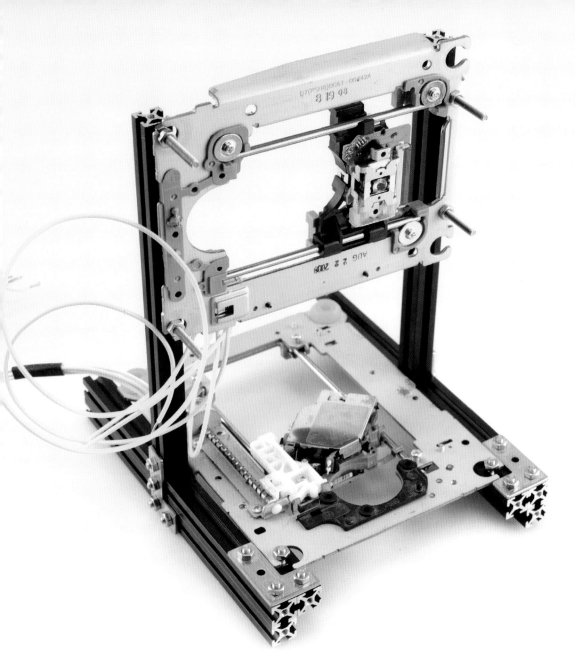

9 We are now going to mount both halves of the CNC machine. Put a square bolt through each of the remaining L bracket holes and slide the top part of the frame onto this. Once it is in place, secure each side with a nut and tighten using the nut driver. The entire assembly needs to be very rigid, so check the alignment of everything and make sure all is tight and secure.

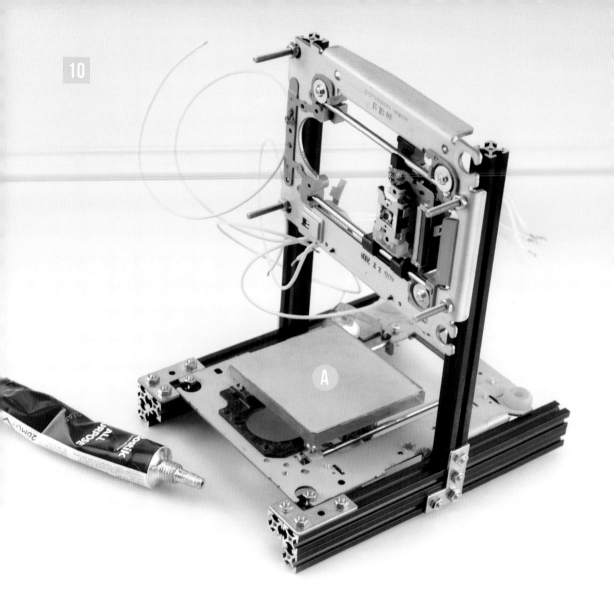

10 Using superglue, stick a small piece of acrylic (A) to the moving part of the Y axis. Don't worry if you don't have any acrylic – any small thick piece of some scrap material to act as a bed will do just as well.

11 We now need to mount a pen to the top drive. To set the 'Z' height (as we have no independent control of this), put a small piece of paper on the bed and remove the cap from the pen. Position the pen so that it just touches the paper and begins to make a mark. Attach the pen to the X axis with superglue, so that the pen is held in exactly this position.

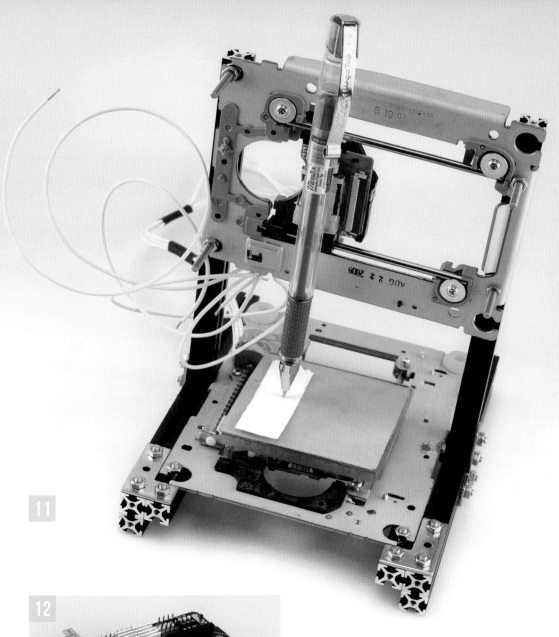

11

12

12 It's time to install the electronics into our CNC machine. First mount the dual bipolar stepper motor controller shield onto the Arduino Uno (A). The sprung terminals should face towards the USB connector on the Arduino Uno.

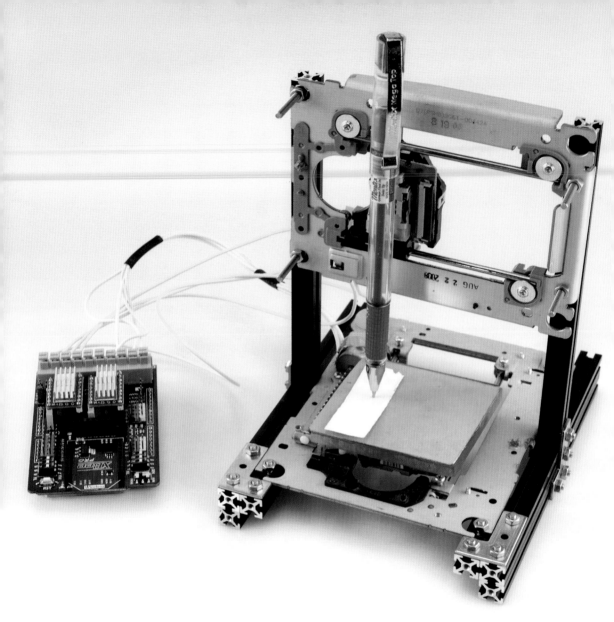

13 Attach the motor wires to the dual bipolar stepper motor controller; the X axis (top drive) wires should go to the terminals marked for the X axis and the Y axis (bottom drive) wires should go to the terminals marked for the Y axis. Remember that we marked the pairs of windings near the beginning? This is where that step matters! One pair should go to the terminals marked 1A and 1B and the other pair of wires should go to 2A and 2B. If you get these wrong, the motor will not move.

Never unplug the stepper motors from the driver when it is powered, this can damage the controller board.

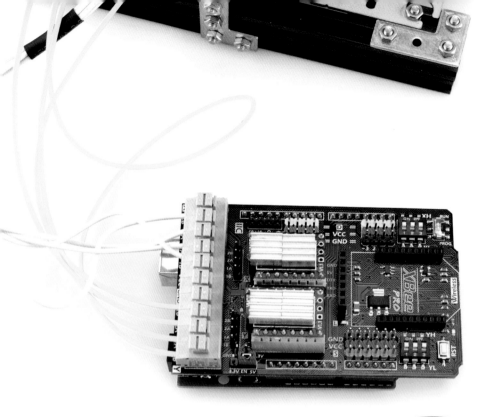

14 I mentioned earlier that the controller board does the fancy pulsing required to control the stepper motors. The firmware put together for this machine uses something known as micro-stepping to make the stepper motor move a tiny amount on each pulse (step). As we increase the amount of micro-stepping, more steps are needed for the motor to move the axis the same physical distance. I've already configured this for you in the firmware, but this means we need to make sure that the CNC controller has been set to the right micro-stepping mode. To do this, push the three dip switches to the 'on' position; do this for each axis.

We've now completed the physical assembly of our CNC machine.

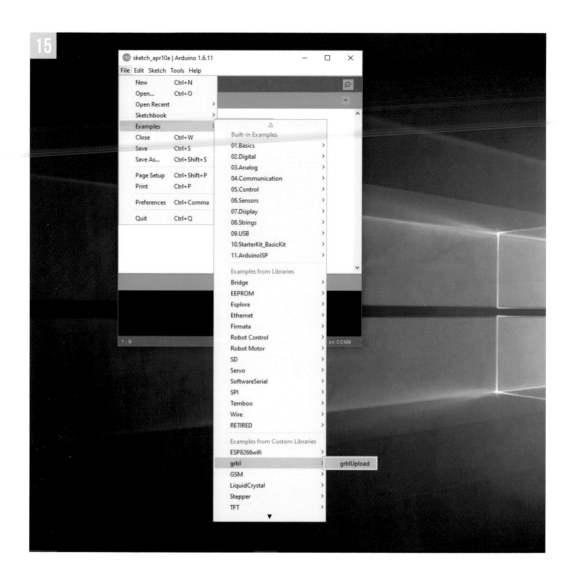

It's now time to upload the software to our Arduino. Download the appropriate zip file from the online website (https://github.com/danielknox/CNC-Machine) and follow the included instructions to extract the required files.

Open the Arduino IDE (Integrated Development Environment) and click the 'File->Examples' tab. Choose 'grbl->grblUpload' from the menu list.

16 We now need to make sure we are compiling the software for the correct type of Arduino board. Click on the 'Tools->Board:' tab and choose the 'Arduino/Genuino Uno' option.

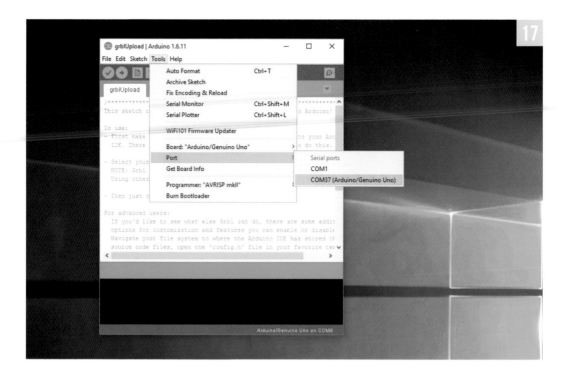

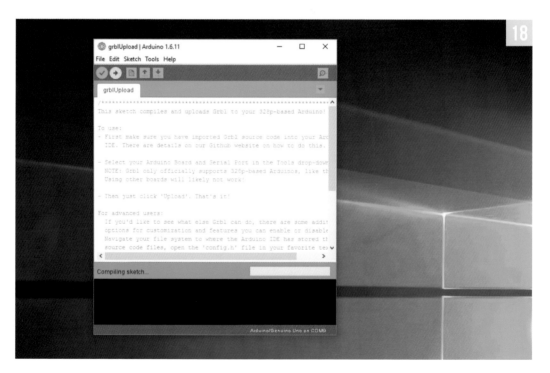

17 Now click 'Tools->Port' and select the 'com' port for the Arduino Uno – you will need the Arduino plugged into your PC for this to be detected. There are probably only going to be a few possible devices in this list; ignore Bluetooth devices on Macs and COM1 on Windows.

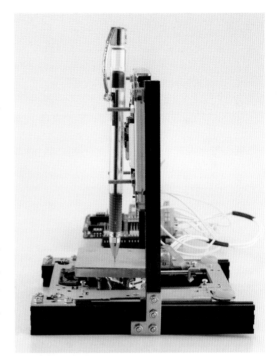

18 We can finally upload the software to the device. To do this click on the icon that looks like a right-facing arrow (if you hover over it you should see an 'Upload' label appear). The Arduino IDE will now compile the program and upload it to the Arduino board for us.

Our CNC machine is now ready to use.

HOW IT WORKS

Plug the 9-volt power supply into the terminals marked 'VCC and GND' on the motor shield, VCC is for the positive cable and GND is for the negative lead. You must make sure you get the polarity right before plugging it into the shield. I like using a multimeter in 'volts' mode to confirm this. Put one probe on each wire and read the voltage on the meter, if there's a little negative symbol, you probably have the wires around the wrong way.

Using CNC machines is a bit more complicated than working with our other robots, so I've put together a little guide on the same page from which you downloaded the software; follow this and before long you will be confident in CNC control.

MARS ROVER

One of the greatest benefits that robots provide is their ability to work in places where it is too dangerous or difficult for humans to function. This can be at the bottom of the sea, inside nuclear reactors, or even on the surface of other planets.

COMPONENTS

8 equal right-angle MakerBeam bracket
6 45-degree-angle MakerBeam bracket
M3 square head set bolts x 6mm (pack of 250)
M3 square head set bolts x 25mm (pack of 25)
4 M3 hex brass standoff spaces
M3 stainless steel plain nuts (pack of 250)
M3 washers (pack of 10)
6 MakerBeam – 100mm (4in) long black anodized beam
2 MakerBeam – 150mm (6in) long black anodized beam
project box (approx 190 x 110 x 60mm/7½ x 4½ x 2½in)
zip ties
5mm threaded rod – 300mm (12in) length
6 M5 stainless steel plain nuts
DFRobot Romeo BLE all-in-one microcontroller
6 6-volt geared robot motors with wheels
22 AWG spool of stranded wire
NiMH 9.6-volt 1800mAh 8 cell AA battery pack
NiMH battery charger with Tamiya mini battery connector
Tamiya mini battery connector
(*see* page 150 for solar panel components)

TOOLS

soldering iron
lead-free solder
drill vice
snips
3mm nut driver (5.5)
wire strippers
rotary multi-tool
cordless drill with 5mm and 3mm drill bits
marker pen
mobile phone or tablet with wifi
superglue
retractable craft knife (only required for solar panel)

SNAPSHOT

The Mars Rover, Sojourner, is one of my all-time favourite robots. It was exploring Mars for 83 sols (as days on Mars are known) while I was still at school learning how to build machines. I've always wanted to build a spiritual replica of Sojourner and now have finally had the chance to do so. Our Sojourner won't be exploring remote planets anytime soon, but it has a similar rocker-bogie suspension system to allow it to traverse tough terrain; it also has a controller that allows easy expansion of the robot with various sensors.

1 We first need to resize one of the holes in each of the right-angled brackets to allow us to put the 5mm threaded rod through it. To do this, grip the bracket inside a vice and use a 5mm drill bit and cordless drill to expand the shared middle hole. Try to get the drill as straight as possible when doing this. Repeat this for all of the right-angled brackets.

2 We are now going to begin assembling the rocker-bogie using aluminium extrusion. This material is great as it comes in various standard sizes and can be cut to length; this allows the easy creation of frames. The downside of aluminium extrusion is that it requires the use of special bolts; the bolt slides into the extrusion and we then secure it in place by using a nut driver to tighten a nut. The particular extrusion we are using here is known as MakerBeam, which gives us a tiny 10mm x 10mm profile. Begin to assemble the first bogie by using two 100mm (4in)

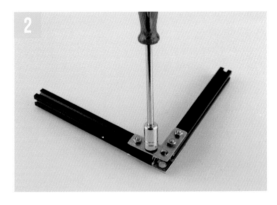

lengths, 6mm M3 square head bolts, and two right-angled brackets. Slot the required square bolts in first and then attach both right-angled brackets at the same time.

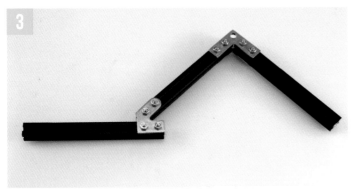

3 Now take a third piece of 100mm (4in) extrusion and use a 45-degree-angle bracket to connect it to our previous frame. We only need one 45-degree-angle bracket for this connection.

4 We are now going to assemble the second part of the bogie; for this, we require two right-angled brackets, two 45-degree-angle brackets, two washers, a longer bolt and a 150mm (6in) piece of extrusion. Put two right-angled brackets, one on each side of the 150mm piece of extrusion. Connect the 45-degree-angle brackets to the end of the part that we assembled in step 2. Finally use a longer bolt to attach the two parts together; include a washer to help the part to pivot freely.

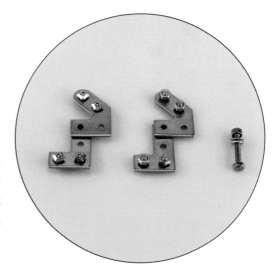

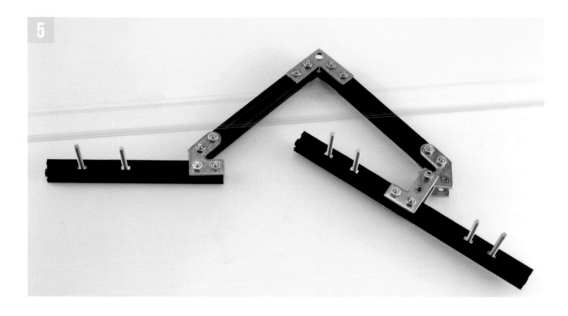

5 We now attach three of the 6-volt motors to the frame. Slot 25mm (1in) square bolts onto the frame; two bolts for each motor.

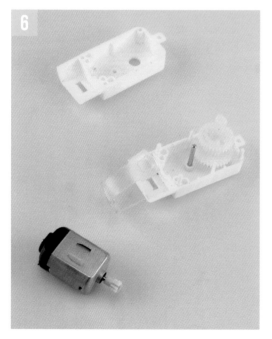

6 The middle motor will not be powered, but left unmodified the gears inside the motor will increase the friction that the other motors will have to overcome on each rocker-bogie. To fix this, open one of the motors and remove the internal gears that connect the drive axle to the motor. Reassemble the motor and you should find that the axle now spins freely.

7 Slot the motors onto each of the 25mm bolts and tighten them in place with nuts. The free-running motor that we made in step 6 should be in the middle. Position the motors so that their electrical terminals face inwards towards the extrusion.

8 Solder a length of wire onto each of the terminals of the powered motors (not the middle one). It needs to be long enough to run up the length of each extrusion with roughly 15cm (6in) of wire to spare.

BRAINWAVE
Use differently coloured wires to help identify which wire runs to which motor terminal.

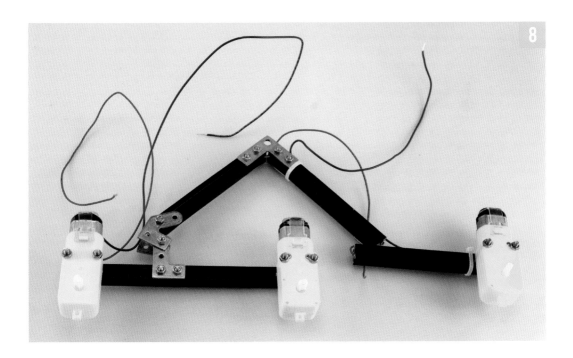

9 Repeat steps 2–8 to produce a second rocker-bogie. This second rocker must mirror the one that we have just produced. Fix the wheels in position. Once completed, put the rocker-bogies to one side.

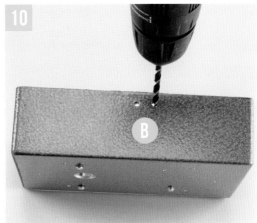

10 We are now going to modify the project box to allow us to install the controller board and the rocker-bogies on it. First mark the centre of the project box, roughly 10mm (¼in) from the top. Mark a second hole adjacent to this (about 15mm/½in away). Drill both holes using a 5mm drill bit in a cordless drill (A). Repeat this step on the opposite side of the box (B). We want the middle holes to be closely aligned as the axle will run through them.

11 Flip the project box over and place the controller board on top. Mark out the location of each of the mounting holes onto the project box. Use the cordless drill and a 3mm drill bit to drill out each of the mounting holes.

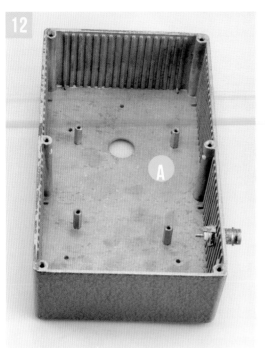

12 Flip the project box back over and install the hex standoff spaces – the thread of the spacer should go through the project box and be tightened in place with a nut (A). Once the spacers are in, insert the controller board and secure this in place with screws (B). Insert the bare wire ends of the Tamiya mini battery connector into the motor power screw terminals.

13 Insert the 5mm threaded rod through the axle holes that we made in the project box. Tighten an M5 nut on each side.

14 Install each of the rocker-bogies onto the axle. Secure the rocker-bogies in place with two M5 nuts on each side. Thread the motor wires through the spare holes in the project box and insert two wires into each of the screw terminal blocks for the motors (M1 and M2 will each drive two motors). Finally, secure the wire lengths to the extrusion using zip-ties.

15 Use a rotary multi-tool with a cut-off blade to trim the excess threaded rod.

MAKE A FAKE SOLAR PANEL

This step is optional and only required if you want to create a fake solar panel, it is not required for the function of the Mars Rover! It also requires access to a special piece of machinery, a laser cutter, but many school workshops have one of these!

COMPONENTS

3mm translucent piece of blue acrylic (600 x 400mm/24 x 16in sheet) black vinyl (300 x 300mm/12 x 12in sheet)

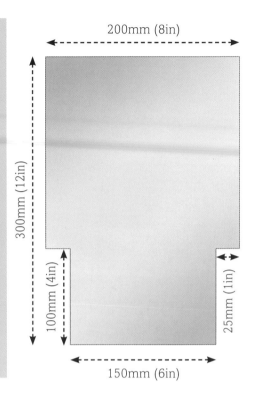

200mm (8in)

300mm (12in)

100mm (4in)

25mm (1in)

150mm (6in)

16 If you are a school pupil, ask your teacher to help to cut the following DXF file (https://github.com/danielknox/Mars_Rover) out of 3mm (⅛ inch) translucent dark blue acrylic; the file is in millimeters. Next remove the adhesive backing off a piece of black vinyl and stick it to one side of the acrylic panel. Normally we want to avoid trapping air bubbles, but in this case, it will add to the effect – just pop the larger ones with a small needle/point of the retractable knife. Cut the excess vinyl away using the knife and then score lines into the top of the acrylic. Finally, glue the acrylic panel onto the lid of the project box.

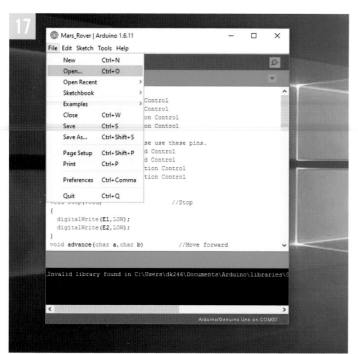

17 We now need to upload the required software to our Arduino based board (the DFRobot Romeo). I've already written the code for the robot – we just need to compile and upload it. Download the appropriate zip file from the online website (https://github.com/danielknox/Mars_Rover) and follow the included instructions to extract the required files. Open the Arduino IDE and click the 'File->Open…' tab.

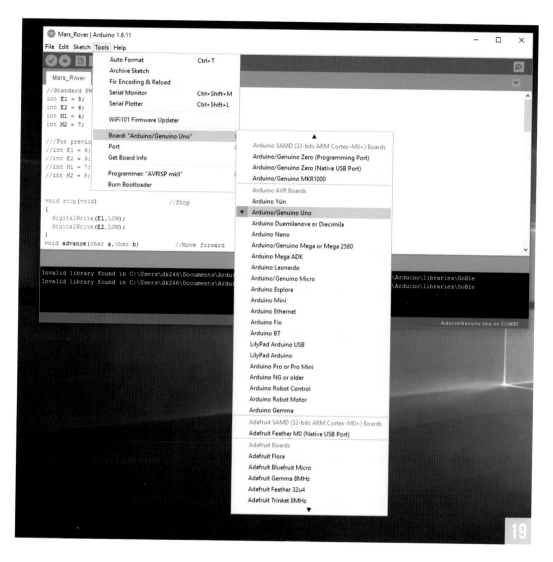

18 Browse to the 'Mars_Rover.ino' file and click 'Open'.

19 We now need to make sure we are compiling the software for the correct type of Arduino based board. Click on the 'Tools->Board:' tab and choose the 'Arduino/Genuino Uno' option.

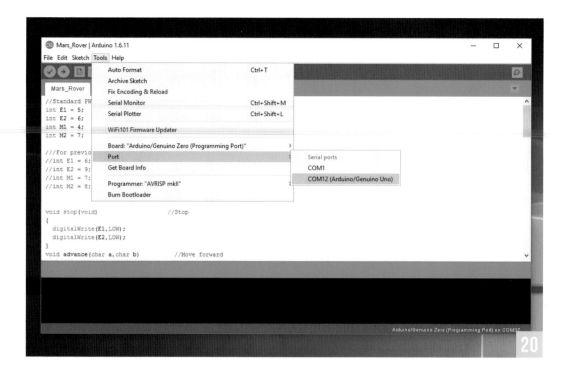

20 Now click 'Tools->Port' and select the 'Com' port for the Arduino based board – you will need the DFRobot Romeo plugged into your PC for this to be detected. There are probably only going to be a few possible devices in this list, ignore Bluetooth devices on Macs and COM1 on Windows.

21 We can finally upload the software to the device. To do this click on the icon that looks like a right-facing arrow (if you hover over it you should see an 'Upload' label appear). The Arduino IDE will now compile the program and upload it to the DFRobot Romeo board for us.

HOW IT WORKS

We are now ready to power up our Mars Rover robot. First charge the NiMH battery using an appropriate battery charger. Once it is ready, plug the NiMH battery into the robot. You should now be able to connect to it using a mobile phone or tablet device. If you are on an Android device, install the app 'Bluno Remote', on IOS select the 'GoBLE' remote controller app.

Once installed, pair your device with the Mars Rover using Bluetooth LE – the pin is '000000'. You should now be able to start the appropriate app and control your Mars Rover with it. If the wheels are spinning the wrong way (so the robot isn't moving), swap the wires around – you want the wheels on each side to rotate in the same direction.

If you do this, make sure you unplug your robot before swapping around the wires. You should be able to operate the robot for around 15 minutes; after that (or once you are finished) unplug the NiMH battery from the robot and recharge it. Don't forget it and leave it plugged in or you will drain the battery.

RESOURCES

The electrical and hardware components used to make the projects in this book are all easily sourced online or at electrical and DIY stores. All the other items that you need will be available at local stores.

USEFUL RESOURCES:

BBC micro:bit – http://microbit.org/,
https://makecode.microbit.org/#lang=en
CNC Machine files –
https://github.com/danielknox/CNC_Machine
Robot Warrior files –
https://github.com/danielknox/Robot_Warrior
Mars Rover files – https://github.com/danielknox/Mars_Rover
electronics & engineering – https://learn.adafruit.com/

COMPONENT SUPPLIERS:

3-volt motor – eBay, Adafruit (US)
4.8V 128mAh ultra-miniature 1/4AAA NiMH battery with
Futaba connector – eBay
6-volt geared robot motors with wheels – eBay, Amazon (US)
9V power supply – eBay, Amazon (US)
22 AWG wire (Stranded) – Maplin, Pimoroni, Adafruit (US)
acrylic, all sizes– theplasticpeople, Amazon (US)
AA batteries – convenience store/supermarket
AA battery holder – eBay, Adafruit (US)
AAA batteries – convenience store/supermarket
AAA battery holder with wires – eBay, Adafruit (US)
Arduino Uno – Maplin, Pimoroni, Adafruit (US)
BBC micro:bit – Maplin, Pimoroni, Adafruit (US)
BBC micro:bit AAA switched battery holder – Maplin,
Pimoroni, Adafruit (US)
CDROM/DVD (old) drives – eBay, boot fair/yard sale
continuous rotation servos – Amazon, eBay, Adafruit (US)
DFRobot dual bipolar stepper motor controller for Arduino –
Robotstore, DFRobot (US)
DFRobot Romeo BLE all-in-one microcontroller – Robotstore,
DFRobot (US)
dog tag blanks – eBay, military surplus store
double sided adhesive foam pad, roll of – DIY/craft store
enamelled craft wire – craft store
'fantasy' film – craft store/eBay
HK-T6A V2 transmitter – HobbyKing

HK-T6A V2 receiver – Hobbyking
insulation tape roll – DIY store
lead-free solder – Maplin, Pimoroni, Adafruit (US)
M3 Hex brass standoff spaces – DIY store, Metric Machine
Screws (US)
M3 stainless steel bolts, plain nuts etc – DIY store, Metric
Machine Screws (US)
M3 square head bolts – technobotsonline, Amazon (US)
M5 stainless steel plain nuts – DIY store, Metric Machine
Screws (US)
Magician robot chassis kit – eBay
MakerBeam angled brackets – technobotsonline, Amazon (US)
MakerBeam, black anodised – technobotsonline, Amazon (US)
metal tin (e.g. Altoids/'survival' tin) – convenience store/eBay
mobile phone vibrator motor – eBay, Pimoroni, Adafruit (US)
nails – DIY store
NiMH 9.6V 1800mAh 8 cell AA battery pack – eBay
NiMH battery charger with Tamiya battery connector – eBay
PIR sensor – eBay, Adafruit (US)
project box – Maplin, Amazon (US)
ring electrical crimp terminals (Red, 3.2mm) – DIY store
servo – eBay, Adafruit (US)
servo tester – Hobbyking
soil hygrometer sensor for Arduino – eBay
SPDT momentary switch with long lever – eBay, Adafruit (US)
Tamiya mini battery connector – eBay
terminal strip – DIY store (electrical aisle)
threaded rod – DIY store
toggle switch (large) – eBay, Adafruit (US)
vinyl sheet, Black – Amazon
wood (sheet of, 6mm) – DIY store
zip ties – DIY store

LASER CUTTING SERVICES:

Razor Lab – http://www.razorlab.co.uk (UK)
Ponoko – https://www.ponoko.com (US)

INDEX

A

AA battery holder 37, 40, 43–47,
   with switch 4.5-volt 12, 67, 70–76
   with wires, switchable 12, 49, 54–55
AAA battery holder
   for BBC micro:bit, switched 12, 67,
     70, 89, 92, 94
   with wires 12, 19, 20–23, 25–28, 31–35
acrylic 114, 127, 134, 152, 158
Allen keys 11, 67, 79, 99
Arduino 9, 13, 89, 127, 138–141,
   154–157, 158
   Uno 127, 135, 141, 155
Avatar 58–65

B

Basic Bots 6, 16–55
basic techniques 15
battery 47, 54, 71, 72, 89, 99, 113
   box, modified 79–84, 99, 103–104
   NiMH, and accessories 12, 113–125,
     143, 150, 157
   pack for BBC micro:bit 65, 77, 87, 96,
     103, 109
BBC micro:bit 7, 12, 13, 56–109
Bluetooth LE 67, 75, 95, 141, 157
bolts 14, 38, 49, 52, 54, 73, 84, 89,
   92, 103
   see also M3 square head set
breakout board/shield 96–97
Bristle Bot 6, 18–23
Butterfly Bot 6, 30–35

C

cardboard 8, 19–23, 26, 37, 79, 82, 113
Catapult Bot 78–87
CDROM/DVD drive 126–141
chassis (*see* Magician robot)
clothes peg 49, 51–55
CNC Writer 9, 126–141
combat, robot 112–125
components guide 12–14
computer/laptop 11, 58–60, 77, 87,
   120–125, 127, 138–141, 154–157

computer numerical control 9, 126–141
continuous rotation servo (*see* motor)
credit card/gift card 113–117
crimping tool 11, 67, 69, 73, 79, 89,
   93, 99
cut-off disc 11, 33, 68, 113
cutting mat 25, 27

D

DFRobot dual bipolar stepper motor
   controller 14, 127, 158
DFRobot Romeo BLE all-in-one
   microcontroller 14, 143, 157, 158
diagram 39, 84, 97, 101, 114, 115, 118,
   120
dog tag blank 98–109
drawing 6, 24–29, 48–55
drill
   bits 11, 49–51, 91, 143–144, 149
   cordless 11, 143–144, 149
drilling vice 11, 31, 33, 49, 50, 67–68, 79,
   89, 91, 113, 116, 143–144
drinks cup, plastic 24–29
DVD drive (*see* CDROM)

E

elastic band 78–87

F

'fantasy' film 30–35

G

Garden Guardian 88–97
glue
   mouldable, Sugru 99, 101
   PVA 31, 34
   stick 25–29
   superglue 67, 72, 79, 82, 85, 89, 94,
     101–104, 127, 134, 143, 152
glue gun 10, 15, 19, 22, 25–26, 28–29,
   31, 35, 37, 44, 47, 49, 51, 89, 94, 113,
   115–118
googly eyes 19, 22, 25–26, 31, 35, 67,
   72, 89, 94

H

helping hands 11, 31–32, 37, 40, 99–101
Hex
   keys 11
   standoff spaces 14, 143, 150, 158
Hobby King HK-T6A V2 receiver/
   transmitter 113, 121, 123, 158

J

JavaScript Blocks Editor 60, 61, 65, 95

K

knife, retractable 10, 19, 20, 25, 27, 37,
   45, 143, 152

L

lighter, gas 31, 99, 101
lollipop stick 49, 50–52, 79–82

M

M3
   Hex brass standoff spaces 11, 143,
     150, 158
   square head set bolts 127, 130, 133,
     143–147, 157
   stainless steel plain nuts 127, 130,
     143
   washers 143, 145
M5 nuts 143, 150–151, 158
Magician robot chassis kit 36–47
MakerBeam 14, 126–141
MakerBeam brackets 14, 126–141
Mars Rover 9, 142–157
metal tin 14, 67–68, 72–73, 89–92, 94,
   97, 158
motor
   3-volt 13, 19–20, 25, 27, 158
   6-volt geared robot motor with
     wheels 13, 37–38, 49, 51, 53, 113,
     143, 146, 158
   9g servo 13, 113, 116, 118
   continuous rotation servo 13, 67–68,
     75, 79, 113–115, 118, 120, 158
   servo 13, 79, 83–85, 87, 99, 100–109

stepper (*see* DFRobot)
mobile phone, vibrator 13, 31, 35, 158
multimeter 10, 89, 93, 127, 129, 141
multi-tool, rotary 11, 31, 49, 50, 53, 67–68, 79, 83, 89, 91, 92, 113, 116, 143, 151

N

nail 25, 27, 158
nail brush 19, 21, 22, 23
needle file 67, 89, 92,
NiMH battery and accessories 12, 113–114, 118, 125, 143, 157, 158
nut driver 10, 127, 130, 133, 143, 144
nuts 14, 38, 49–50, 52, 54, 67, 73, 79, 84, 89, 92, 99, 103, 143, 150–151, 158

P

paper 25, 28, 49, 54, 134
paperclip 19, 21, 22, 37, 45, 47, 53
pen
    ballpoint 89, 94, 134
    marker 10, 19, 21, 25, 29, 49, 50, 53–55, 67–68, 79, 90–92, 113, 127, 143
pipe cleaners 25, 26
PIR sensor 14, 79, 83, 84, 87, 158
pliers, engineer 10, 19, 21, 31–33, 49, 55, 89, 93
precision snips 10, 83, 127, 143
programming 7, 9, 58–65, 67, 75–76, 85–87, 105–109, 138–141, 154–157
project box 14, 79, 83, 85, 99, 143, 150–152, 158
power supply, 9v 26–141

R

ring electrical crimp terminals 13, 67, 69, 73, 79, 83, 89, 93, 99
Robo Roach 6, 36–47
Robo Warrior 8, 112–125

rotary multi-tool 15, 33, 50–53, 68, 83, 91–92, 116, 151
rover vehicle 9, 142–157
rubber band (*see* elastic band)

S

safety glasses 11, 31, 33, 49, 50, 53, 89, 91–92
sandpaper 25, 28, 31, 35, 94
scissors 19, 21, 31, 34
screwdriver
    cross-head 10, 37, 49, 53
    flat-head 10, 19–20, 49, 52, 54, 70, 89, 96, 113, 118, 127
screws 14, 20, 67, 79, 99, 128, 150
Scuttle Bot 7, 66–77
servo (*see* motor)
servo tester 13, 79, 83, 99–100, 158
Simple Robots 7, 56–109
Smart Makes 8–9, 110–157
snips (*see* precision snips)
socket head 14, 67, 79, 99
software 6, 7, 9, 90, 96, 121, 123–124, 138–139, 141, 154–155, 157
soil hygrometer sensor 88–97, 158
solar panel 143, 152–153

soldering iron 11, 15, 20, 27, 31–32, 37, 39, 40, 42–44, 49, 55, 67, 70–71, 127, 129, 130, 143, 147, 158
SPDT momentary switch with long lever 13, 37, 39, 40, 42–44
Spiro Bot 6, 48–55
Squibble Bot 6, 24–29

T

Tamiya battery and accessories 143, 150, 158
tape
    double-sided adhesive foam 67, 72
    insulation 73, 79, 84, 92, 99, 104, 158
    masking 49, 50, 54

techniques, basic 15
terminal strip 13, 19, 20, 158
threaded rod 14, 143, 150–151, 158
toggle switch 89, 94, 158
tools guide 10–11
toothbrush 18–23
transmitter 8, (*see also* Hobby King)

U

USB cable 59, 113, 121

V

vice, drilling (*see* drilling vice)
vinyl, sheet 152, 158

W

Walking Robot 7, 98–109
washers 14, 49, 52
    spring 49, 52
WiFi 143
wire
    craft, enamelled 31, 33
    stranded 20 AWG 36–47, 48–55
    stranded 22 AWG 126–141, 142–157
wire strippers 11, 31, 37, 42, 67, 69–70, 79, 83, 89, 93, 99, 127, 143
wheels 13, 37–38, 46, 49, 51–55, 113, 119, 143, 148, 157, 158
wood 49, 53–55, 114, 158